"This engaging volume is an account of the human grasp, measurement, and use of time. It is comprehensive and detailed, yet enthralling in the way the history is told and the technical aspects explained."
—HUGH DOWNS

Our present calendar predates the invention of the telescope, the mechanical clock, and the concept of zero—and its development is one of the great untold stories of science and history. Now, David Ewing Duncan leads us on an extraordinary journey through man's reckoning of time, from the earliest calendars through our struggles with the digital "millennium bug." We travel from Stonehenge to Giza, from Mayan observatories to the atomic clock in Washington; we visit a host of ancient cultures and meet a cast of historic personages and giants of science. Here is a fresh, stimulating volume that answers—and raises—a host of facinating questions about the nature of human timekeeping and the majestic historical forces that have produced the miracle of the calendar.

"Veteran science and travel writer Duncan has assembled lively history— dating back 13,000 years to the first known timetable—of the attempt to follow our exact place in the whirl of days, lunar cycles, seasons, and years."
—OUTSIDE MAGAZINE

"Duncan is a master at weaving together various threads and anecdotes . . . He sketches out fascinating characters . . . By using a tiny thread to guide us through ages, cultures and religions, CALENDAR is a fascinating exploration of the history of ideas— and a chance to reflect on the exact nature of the little grid of boxes that rules so much of our lives."
—SAN JOSE MERCURY NEWS

Also by David Ewing Duncan

RESIDENTS: THE PERILS AND PROMISE OF EDUCATING
YOUNG DOCTORS
HERNANDO DE SOTO: A SAVAGE QUEST IN THE AMERICAS
FROM CAPE TO CAIRO: AN AFRICAN ODYSSEY
PEDALING THE ENDS OF THE EARTH

Calendar

Humanity's Epic Struggle to Determine a True and Accurate Year

David Ewing Duncan

AN AVON BOOK

AVON BOOKS, INC.
1350 Avenue of the Americas
New York, New York 10019

Copyright © 1998 by David Ewing Duncan
Cover Illustration by Christine Van Bree
Inside back cover author photograph © Kit Morris Photography
Interior design by Kellan Peck
Time line illustration by Myles Sprinzen
Published by arrangement with the author
ISBN: 0-380-79324-5
www.avonbooks.com/bard

Library of Congress Cataloging in Publication Data:

Duncan, David Ewing.
Calendar : humanity's epic struggle to determine a true and
accurate year / by David Ewing Duncan.—1st ed.
    p.   cm.
  Includes bibliographical references (p.   ) and index.
1. Calendar—History   I. Title.
CE6.D86   1998
529'.3'09—dc21   98-10434
CIP

First Bard Trade Paperback Printing: June 1999
First Bard Hardcover Printing: July 1998

BARD TRADEMARK REG. U.S. PAT. OFF. AND IN OTHER COUNTRIES, MARCA REGISTRADA, HECHO EN U.S.A.

Printed in the U.S.A.

OPM   10 9 8 7 6 5 4 3 2 1

To Sander, Danielle,
and Alexander

and thanks to Stephen

Calendar Index

Length of the (tropical) year in 2000 A.D.: 365 days, 5 hours, 48 minutes, 45 seconds

Time that the year has slowed since 1 A.D.: 10 seconds

Average decrease in the year due to a gradual slowing of the earth's rotation: ½ second per century

Lunar Month: 29 days, 12 hours, 44 minutes, 2.9 seconds

The earliest known date: 4236 B.C., the founding of the Egyptian calendar

Ancient Egyptian year: 365¼ days

Early Chinese year: 354 days (lunar year) with days added at intervals to keep the Chinese lunar calendar aligned with the seasons

Early Greek year: 354 days, with days added

Jewish year: 354 days, with days added

Early Roman year: 304 days, amended in 700 B.C. to 355 days

The year according to Julius Caesar (the Julian calendar): 365¼ days

Date Caesar changed Roman year to Julian calendar: January 1, 46 B.C.

Amount of time the old Roman calendar was misaligned with the solar year as designated by Caesar: 80 days

Total length of 46 B.C., known as the "Year of Confusion," after adding 80 days: 445 days

The year as amended by Pope Gregory XIII (the Gregorian calendar): 365 days, 5 hours, 48 minutes, 20 seconds

Date Pope Gregory reformed the calendar: 1582

Length of time the Julian calendar overestimates the solar year per year, as determined by Pope Gregory: 11 minutes, 14 seconds

Number of days Pope Gregory removed to correct the calendar's drift: 10

Dates Gregory eliminated by papal bull to realign his calendar with the solar year: October 5–14, 1582

Dates most Catholic countries accepted the Gregorian calendar: 1582–1584

Date Protestant Germany accepted the Gregorian calendar: partial acceptance in 1700, full acceptance in 1775

Date Great Britain (and the American colonies) accepted the Gregorian calendar: 1752

Length of time eliminated by the British Parliament to realign the old calendar (Julian) with the Gregorian calendar: 11 days

Dates Parliament eliminated: September 3–13, 1752

Date Japan accepted the Gregorian calendar: 1873

Date Russia accepted the Gregorian calendar: 1917 (and again in 1940)

Date China accepted the Gregorian calendar: 1949

Date the Eastern Orthodox Church last voted to reject the Gregorian calendar and retain the Julian calendar: 1971

Length of time the Gregorian calendar is off from the true solar year: 25.96768 seconds per year

Length of time the Gregorian calendar has become misaligned over the 414 years since Gregory's reform in 1582: 2 hours, 59 minutes, 12 seconds

Year in which the Gregorian calendar will be one day ahead of the true solar year: A.D. 4909

Year that Atomic Time replaced Earth Time as the world's official time standard: 1972

The year as measured in oscillations of atomic cesium: 290,091,200,500,000,000

The Year 2000 Will Be:

1997 according to Christ's actual birth circa 4 B.C.
2753 according to the old Roman calendar
2749 according to the ancient Babylonian calendar
6236 according to the first Egyptian calendar
5760 according to the Jewish calendar
1420 according to the Moslem calendar
1378 according to the Persian calendar
1716 according to the Coptic calendar
2544 according to the Buddhist calendar
5119 in the current Maya great cycle
  208 according to the calendar of the French Revolution
the year of the **DRAGON** according to the Chinese calendar

Time is the greatest innovator.
—FRANCIS BACON, 1625

Contents

Prelude: A Net Cast Over Time

The . . . silent, never-resting thing called time, rolling, rushing on, swift, silent, like an all-embracing ocean tide . . . this is forever very literally a miracle; a thing to strike us dumb.
—THOMAS CARLYLE, 1840

Not long ago I met a well-known surgeon dying in a hospital in Richmond, Virginia. He was a distressingly emaciated figure, his face a mask of skin over his skull, his hands a pale shade of purple from weeks of intravenous needles. Yet his voice remained deep and powerful, his eyes lively. When a friend asked how long he was going to be in the hospital this time, the surgeon said he didn't know, that time was becoming irrelevant to him. "It's ironic," he said, smiling weakly. "I lived by the calendar for sixty years. Beepers, schedules—these things ruled my life. Now I have no idea what day it is, and this doesn't bother me. It's as if I am floating," he said, leaning back on crisp hospital sheets and almost whispering the words.

Our obsession with measuring time is itself timeless. After self-awareness, it may be our most distinctive trait as a species, since undoubtedly one of the first things we became self-aware about was our own mortality—the fact that we live and die in a set period of time.

Yet even in an age of measuring femtoseconds* and star clusters 11 billion light-years away, time defies true objective measurement. It

---

*A femtosecond is one quadrillionth of a second.

can seem to go slow and even stall out at certain moments only to brashly and breathlessly rush forward at others. Time can be wasted, kept, saved, spent, killed, lost, and longed for. To the Nuer herdsmen of southern Sudan, time is *tot* and *mai*, wet and dry, depending on the season. For Hesiod, the ancient Greek poet, time is harvesting cereals in the mouth when the cuckoo sings, and a low sex drive for men during the late summer, when "goats are at their fattest and the wine tastes best."

Consider the geometry of how we measure time. It can be divided into circle time and square time: clock time and calendar time. Clock time chases itself like Ouroboros, the hands or flashing numbers returning to the place where they started in a progression that has no beginning or end. It will continue in its cycle whether or not people are around to watch the hands and glowing numbers. In contrast, calendar time is made up of small boxes that contain everything that happens in a day, but no more. And when that day is over, you cannot return to that box again. Calendar time has a past, present, and future, ultimately ending in death when the little boxes run out.

Still, in modern times we take the mechanism of the calendar for granted, as we do breathing and the force of gravity. Passing through years, months, weeks, hours, minutes, and seconds we seldom think about where these things came from, or why we have chosen to divide time one way and not another.

It has not always been so. For thousands of years the effort to measure time and to create a workable calendar was one of the great struggles of humanity, a conundrum for astronomers, mathematicians, priests, kings, and anyone else who needed to count the days until the next harvest, to calculate when taxes were due, or to figure out the exact moment a sacrifice should be made to appease an angry god. A case can be made that science itself was first sparked by a human compulsion to comprehend the passing of time, to wrestle down the forward motion of life and impose on it some sense of order.

The effort to organize and control time continues unabated today. It is one of humankind's major collective efforts as we hedge our future and try to comprehend the past. In the stock market an investor sells a microchip stock short or long based on a reading of the company's history. In river valleys we build dams and levees to prepare for 10-, 50-, and 100-year floods. We celebrate Easter, Passover, and Ramadan on prearranged dates just as our ancestors did centuries ago, and we expect our children will for centuries more to come.

We are a people of the calendar. Forward- and backward-looking, we are uncomfortable with the present in a way that our ancestors who tilled fields and lived and died according to the great cycles of nature would never have comprehended.

What are you doing at one o'clock tomorrow? Can you book me on the 2:06 flight to Memphis next Thursday? When will the inventory ship? Ten-nine-eight-seven-six-five-four-three-two-one-zero: *blastoff!*

Holding the surgeon's wasted hands in that Richmond hospital, I thought about my schedule for the rest of the day. Meetings, engagements, phone calls to make, a plane to catch to fly back home. I needed to pick up a small present for my eight-year-old, and I had to remember to put gas in my rental car before I turned it in at the airport. In a way I envied the doctor because he could let go and I could not. This is our blessing and our curse: to count the days and weeks and years, to calculate the movements of the sun, moon, and stars, and to capture them all in a grid of small squares that spread out like a net cast over time: thousands of little squares for each lifetime. How this net was woven over the millennia, and why, is the subject of this book.

Calendar

A Lone Genius Proclaims the Truth About Time

The calendar is intolerable to all wisdom, the horror of all astronomy,
and a laughing-stock from a mathematician's point of view.
—ROGER BACON, 1267

Seven centuries ago a sickly English friar dispatched a strident missive to Rome. Addressed to Pope Clement IV, it was an urgent appeal to set right time itself. Calculating that the calendar year was some 11 minutes longer than the actual solar year,* Roger Bacon informed the supreme pontiff that this amounted to an error of an entire day every 125 years, a surplus of time that over the centuries had accumulated by Bacon's era to nine days.† Left unchecked, this drift would eventually shift March to the dead of winter and August to the spring. More horrific in this pious age was Bacon's insistence that Christians were celebrating Easter and every other holy day on the wrong dates,

---

*When I refer to the "year" or the "solar year," I mean the *tropical year* unless otherwise indicated. The tropical year is the year according to the passing of the seasons. Usually this is defined as the length of time between two successive vernal equinoxes. Because this measurement fluctuates from year to year, the tropical year is usually calculated as an average of several years. The tropical year is slightly different from the *sidereal* year, which measures the length of time it takes for the earth to orbit the sun, returning to a starting point according to a fixed point such as a star.
†In this same treatise Bacon elsewhere uses the figure once in every 130 years. The actual error is closer to once every 128 years.

a charge so outrageous in 1267 that Bacon risked being branded a heretic for challenging the veracity of the Catholic Church.

Roger Bacon did not care. One of medieval Europe's most original and curmudgeonly thinkers, he seemed to relish his role as a rebel—first as a master at the University of Paris in the 1240s and then as a priest after he joined the Franciscan order sometime during the 1250s, when he was in his forties. Insatiably curious and always willing to challenge orthodoxy, Bacon devoted his life to pondering what causes a rainbow, diagramming the anatomy of the human eye, and devising a secret formula for gunpowder. Two centuries before Leonardo da Vinci he predicted the invention of the telescope, eyeglasses, airplanes, high-speed engines, self-propelled ships, and motors of enormous power. He drew these conclusions based on the then-radical notion that science offered objective truths regardless of dogma or what was written down in a book.

Bacon's contemporaries were impressed by his intellect but frightened of his ideas. His own monastic brothers at Oxford and Paris may have held him under house arrest. Even worse, they banned him for long periods from writing and teaching, keeping him busy with the mundane chores of the monastery—tending the garden, reciting prayers, scrubbing the floors. Occasionally they punished him by withholding food.

This might have been the end of Roger Bacon's story if not for a sudden interest in his ideas by a man named Guy Le Gros Foulques. In 1265 this former lawyer and advisor to King Louis IX of France became aware of Bacon and contacted him, asking the friar to send him a compendium of his thoughts. Like Bacon, Foulques had joined the priesthood later in life, in 1256, the year his wife died. He then had advanced with breathtaking speed from priest to bishop, archbishop, and cardinal, his position when he approached Bacon. How Foulques heard about the long-cloistered friar is unknown, nor is it clear why this important cardinal was interested in Bacon's ideas, nor if he agreed with them.

Whatever his reasons, Foulques's interest was a dramatic turn of events for Roger Bacon. The long-suffering friar must have felt as if he were finally being allowed to bring fire back to the cave. And if this were not enough, just a few months later Guy Le Gros Foulques was elected supreme pontiff of the Catholic Church, taking the name of Clement IV. This led to a second notice sent to Bacon: a papal mandate delivered in June 1266 to forward the friar's work as soon as possible to St. Peter's in Rome.

Bacon was elated but embarrassed, for after years of persecution by his own religious order, including at times a prohibition on writing, he had nothing complete enough to send to Rome. "My superiors and my brothers," a frustrated Bacon wrote to the pope, "disciplining me with hunger, kept me under close guard and would not permit anyone to come to me, fearing that my writings would be divulged to others than . . . themselves."

Free at last to pursue his ideas, Roger Bacon promised to prepare a manuscript and send it as quickly as possible. For nearly two years he worked feverishly, finally dispatching an epic treatise to Rome in 1267 called *Opus Maius* (Major work). In this book and two others, hand-carried by a faithful servant named John along the sometimes treacherous medieval highway across Europe, Bacon expounds on topics ranging from a study of languages and the geometry of prisms to the geography of the Holy Land.

His diatribe describing the flaws in the calendar falls in a long and rambling chapter on mathematics, in a section where he advocates using the objectivity of numbers and science to expose mistakes. He opens with an announcement that he is bringing up a matter "without which great peril and confusion cannot be avoided," an error conceived out of "ignorance and negligence . . . [that is] contemptible in the sight of God and of holy men . . . The matter I have in mind," he says, "is the correction of the calendar."

Bacon traces flaws in the calendar back to its originator, Julius Caesar, who launched the calendar used by Bacon—and by us today with some modifications—on January 1, 45 B.C. "Julius Caesar, instructed in astronomy, completed the order of the calendar as far as he could in his time," writes Bacon:

> But Julius did not arrive at the true length of the year, which he assumes to be in our calendar 365 days, and one fourth of a day. . . . But it is clearly shown . . . that the length of the solar year is not so great, nay, less. This deficiency is estimated by scientists to be about the one hundred and thirtieth part of one day. Hence at length in 130 years there is one day in excess. If this were taken away the calendar would be correct as far as this fault is concerned. Therefore, since all things that are in the calendar are based on the length of the solar year, they of necessity must be untrustworthy, since they have a wrong basis.

Bacon also condemns a second calendric mistake that comes out of the first. "There is another greater error," writes Bacon, "regarding the determination of the equinoxes and solstices. For . . . the equinoxes and solstices are placed on fixed days. . . . But astronomers are certain that they are not fixed, nay, they ascend in the calendar, as is proved without doubt by tables and instruments."

This second point was critical, Bacon notes, because the spring equinox—astronomically the point between winter and summer at which the sun strikes the equator—is the date used by Christians to determine Easter. According to Church rules, Easter is celebrated on the first Sunday after the first full moon after the spring equinox.* In Bacon's day the equinox was permanently fixed on March 21 by order of the Church, as established by an important Christian council held at Nicaea in Turkey in A.D. 325. But since 325, as Bacon notes, the equinox had been "ascending in the calendar . . . and likewise the solstices and the other equinox" by 1/130 of a day each year, or just over 11 minutes. He set the true date of the equinox for the year he was writing, 1267, on "the third day before the Ides of March," or

---

*The actual calculation of Easter is considerably more complicated than this, but this simplification will suffice for now.

March 12—a nine-day difference. "This fact cannot only the astronomer certify," says Bacon, "but any layman with the eye can perceive it by the falling of the solar ray now higher, now lower, on the wall or other object, as anyone can note."

He calculates that by 1361 the calendar would drop back another whole day, throwing the entire progression of dates and sacred days further into disarray. The friar concludes with an appeal to Clement to embrace the "truth" offered by science, and to fix the mistake:

> Therefore Your Reverence has the power to command it, and you will find men who will apply excellent remedies in this particular, and not only in the aforesaid defects, but in those of the whole calendar. . . . If then this glorious work should be performed in your Holiness' time, one of the greatest, best, and finest things ever attempted in the Church of God would be consummated.

Bacon's own solution was to drop a day from the calendar every 125 years. But he adds a word of warning, noting that "no one has yet given us the true length of the year, with full proof, in which there was no room for doubt"—a reality that would continue to complicate a final solution for the calendar problem for centuries to come.

Roger Bacon was hardly the first to realize the calendar's drift against the solar year. A millennium earlier the Greek astronomer Claudius Ptolemy (c. 100–178) had noted that the calendar year fell short against the true year, though his calculation differed substantially from Bacon's. In *The Almagest,* a work on astronomy widely read (if not fully understood) during the Middle Ages, Ptolemy sets the drift at about one three-hundredth of a day, a slippage of an entire day every 300 years. This amounts to a five-minute shortfall, or a year of 365 days, 5 hours, and 55 minutes, rather than Caesar's year of 365 days and 6 hours (365¼ days). "And this number of days," writes Ptolemy, "can be taken by us as the nearest approximation possible from the observations we have at present." Considering that Ptolemy, like Bacon, had no telescope and believed that the sun revolved around the earth, this calculation was a reasonably close approximation, though less accurate than Bacon's year of 365 days, 5 hours, and 49 minutes.

Between Ptolemy and Bacon scholars in Europe and Asia tinkered with solutions, attempting to refine earlier estimates of the true year— but always falling short (or long). These tinkerers included the great Indian astronomer Aryabhata (476–550); the mathematician Mohammed Ibn Musa al-Khwarizmi (c. 780–850) and others in the Islamic world; and a progression of mostly obscure monks and scholars in the West, the best-known being the Venerable Bede (673–735) of Britain. Using a sundial in a Northumbria monastery, Bede suspected that the solar year was slightly off from the calendar but did not know by how much. In part this was because Europeans after the collapse of Rome either ignored or did not understand complex fractions. They tended to round off anything but a simple fraction such as one quarter or one half.

Other monks who tried, and failed, to calculate the true solar year include Notker the Stammerer, a Swiss priest-scholar who challenged the accuracy of certain saint's days in a treatise written in about 896; the French ecclesiastic Hermann the Lame, who dared to suggest in 1042 that the Church-approved calendar might be misaligned with the heavens; and Reiner of Paderborn, who made his attempt in the 1100s. But none of these computists dared challenge the Church on a matter so fundamental as measuring time.

Then came Roger Bacon, who seized the opening offered him by Clement to plunge into this ancient puzzle. Dismissing with a wave of his quill centuries of reticence by fearful astronomers, Bacon declared that anyone who rejected the truth offered by science was a fool.

Clement's reaction to Bacon's pronouncements and appeals is unknown. On November 29, 1268, the pope suddenly died, probably before he had a chance to read Bacon's just completed opus.

Nothing could have been more disastrous for the friar, who had just accused the Church of ignorance and wrongheadedness and demanded reforms that Vatican officials less sympathetic than Clement might have condemned as heresy. Instead the Holy See did something far more damning: they ignored him. Clement's successor, Gregory X, never mentions Bacon or his books; nor does anyone else at St. Peter's.

But Bacon continued to speak his mind. In 1272 he penned a blistering attack on academics and what he considered the abysmal state of learning. It spared no one, including universities, kings and princes, lawyers, and the papal court. He also began applying his standards for truth and objectivity to the practice of Christianity, joining a small but vibrant movement of monks scattered across Europe who believed that the Church had strayed from the original dictates of Christ by acquiring too much worldly wealth and power.

Ultimately Bacon's radical talk landed him back in serious trouble. In 1277 he was denounced again by his own religious order, which charged him with espousing "suspected novelties." This time they did not merely cloister him: they sent him to prison. According to Franciscan records of his trial, their high council "condemned and reprobated the teaching of Friar Roger Bacon of England," forbidding anyone from studying his work. They also requested that Pope Nicholas III issue a decree commanding that the friar's "dangerous teaching might be completely suppressed."

For the next decade and a half Roger Bacon disappeared. Then in 1292 the elderly friar, now in his late seventies and apparently out of prison, emerged once more to pen yet another firebrand essay—his last. By then, however, the name Roger Bacon was so obscure that this unfinished and unpublished manuscript was noticed by no one. Nor did anyone bother to record the exact date of his death, possibly in that same year.

But Bacon's passion for the truth endured. Centuries later Roger Bacon became a posthumous hero of late Renaissance and Enlightenment thinkers, who were astonished by the modernity of his ideas.

It took another three centuries before Bacon's demands for calendar reform were heeded, when Pope Gregory XIII (1502–1585) finally fixed the calendar in 1582. By then scientists had been openly clamoring for a correction for several decades. Even Copernicus, a generation before Gregory's correction, penned a section about the true length of the year in his *Revolutions of Heavenly Spheres,* published in 1543. This was the same treatise that offered a compelling theory overturning the age-old belief that the sun and the planets revolved around the earth.

Gregory's reform came after he appointed a calendar commission in either 1572 or 1574, led by the Bavarian mathematician Christopher Clavius (1537–1612), one of two quiet heroes of the Gregorian correction. The other was an obscure Italian physician named Aloysius Lilius (1510–1576), who actually devised the solution Gregory issued as a papal bull on February 24, 1582. This came almost exactly 316 years—and two and a half additional lost days—after Roger Bacon's appeal to Clement IV.

Today almost everyone takes the precision of our calendar for granted, unaware of the long threads spooling out from our clocks and watches backward in time, running through virtually every major revolution in human science, all linked to the measurement of time. The thread largely runs through the West, since this is the source of the world's civic calendar, but also casts lines of varying sizes and thickness outward to China, India, Egypt, Arabia, and Mesopotamia. Unwinding backward, it pauses at Clavius and at Bacon; at the rush of knowledge coming from Islam and the East during the Middle Ages; at bloody wars fought over dates after Rome's collapse; and at Rome at its height, when Julius Caesar fell in love with Cleopatra, an affair that gave the West its calendar. It moves back further still to the Egypt of the pharaohs, Babylon, Sumer, and beyond, thousands of years before Roger Bacon penned his treatise to the pope, when an unknown man dressed in reindeer skins and clutching an eagle bone gazed at the sky and got an idea as radical in his day as anything Bacon thought of in his: to use the moon to measure time.

2
Luna: Temptress of Time

He appointed the moon for seasons: the sun knoweth his going down.
—Psalm 104:19, c. 150 B.C.

Some 13,000 years ago, when the southern flank of the great Würm icecap still touched the Baltic Sea, the Dordogne Valley in central France looked more like present-day Alaska than the leafy wine-growing hills of today. Sprawling herds of reindeer, bison, and woolly rhino grazed on tundra and drank the water of bracingly cold streams. From limestone heights saber-toothed tigers scanned the herds as eagles circled slowly thousands of feet up in the chilly air, looking for shrews, mice, and Paleolithic rodents now extinct.

Perched on a small bluff near what is now the village of Le Placard, another creature gazed not at the deer and roiling stream but skyward. A Cro-Magnon version of Roger Bacon, this hairy, reindeer-skin-clad man patiently waited for the moon to rise above the valley. He was about to revolutionize the way he and his people would view time.

For several nights this Stone Age astronomer and time reckoner had been watching the pale orb in the sky wax and wane. He noted that it moved through a series of predictable phases, and that he could count the nights between the moments when it was full, half full, and completely dark. This was useful information for a tribe or clan that wanted to use the silvery light to cook and hunt, or for calculating future events such as the number of full moons between the first freeze of winter and the coming of spring. For the calendar maker himself it was valuable information he could use to impress his family, his mate, and his clan by predicting when the moon would next be full or would disappear, events that even today signal key religious ceremonies and celebrations.

The man at Le Placard was hardly the first to use the moon as a crude clock. But on this particular night this Cro-Magnon did not merely gaze upward and ponder the phases of the earth's satellite. Turning from the sky, he carefully carved a notch into an eagle bone the size of a butter knife, adding it to a series of notches running vertically along the bone. The notches were straight lines with smaller diagonals carved near the bottoms, looking like this: ⅃. The man added his marking that night to what appear to be distinct groupings of similar symbols that change in regular patterns, possibly corresponding to the phases of the moon. The groupings contain seven marks apiece, which is a close approximation of the moon's progression through new, quarter, full, quarter, and back to new. Sometime later the man discarded or lost this eagle bone for archaeologists to find some 13,000 years later in a dig.

Was this one of the first calendars?

Anthropologists say it is possible, that something like the imagined scene above may have happened on a long-ago night in Le Placard. But not all agree. Skeptics insist that the markings on this and other bones are not calendars but decorations or even random scratches— Stone Age doodles or the marks left when ancient hunters sharpened their knives. Yet over the years archaeologists keep finding the same or similar patterns appearing on stones and bones from sites in Africa and Europe.

One bone, dating back to the Dordogne 30,000 years ago, is covered with rounded gouges that seem to represent the moon's course over a two-and-a-half-month period. Another famous image, the 27,000-year-old "Earth Mother of Laussel," shows the carving of

Carved Eagle Bone
Possible Lunar Calendar
Le Placard, c. 11,000 B.C.

what appears to be a pregnant woman holding a horn marked with thirteen notches. Is this supposed to represent a rough approximation of a lunar year? If so, and if the scratches and notches on the bones and stones really are calendars, then how exactly was this information used? We may never know, though I suspect that our calendar, made of little boxes and numbers, would be just as puzzling to the time reckoner of Le Placard and his clan. Still, a link exists between our calendar and theirs. Both represent conscious efforts to organize time by measuring it and writing it down. And both use astronomic phenomena as a timekeeper, though the Cro-Magnons who carved these bones and stones were clearly people of the moon where measuring time was concerned, while we are people of the sun.

It makes sense that this Stone Age calendar maker chose Luna as his inspiration. Alluring and lovely in its silvery dominance of the nighttime sky, the moon seems at first glance to be a perfect clock in its dependable regularity. Roughly every 29½ days it passes through its phases, from new moon to full moon and back again—a steady celestial progression that anyone can see and keep track of. It is also relatively simple to figure out that twelve full cycles of the moon seem to roughly correspond with the seasons, which is how early societies invented the concept of a time span called a year.

Nearly every ancient culture worshiped the moon. Ancient Egyptians called their moon god Khonsu, the Sumerians Nanna. The Greek and Roman goddess of the moon had three faces: in its dark form it was Hecate, waxing it was Artemis (Diana), and full it was Selene (Luna). Even today people celebrate the moon, holding feasts, dances, and solemn rituals when the moon is new. The San of Africa, for instance, chant a prayer: "Hail, hail, young moon!" Eskimos eat a feast of fish, reportedly put out their lamps, and exchange women. Moslems watch for the new moon of Ramadan, Islam's holy month of fasting and sexual abstinence during the day and feasting at night.

A lunar lexicon still draws on the mystery and majesty of this strange orb hanging in the sky. We have *lunatic* and *moonstruck,* which come from legends that sleeping in the moonlight will drive a person insane. We also have *moonshine, blue moons,* Debussy's *Clair de Lune,* and Shakespeare's comparison of the crescent moon to a "silver bow newbent in heaven." Even more profound is the deeply rooted connection of the moon with measurement. *Me-* or *men* is derived from moon, as in the English *meter, menstrual,* and *measure.*

The moon was not the only early clock, but one of several natural cues used by ancient peoples to measure time and to predict events such as winter, seasonal rains, and the harvest. In Siberia the Ugric Ostiak still base their calendar on natural cycles, incorporating them into month names such as Spawning Month, Ducks-and-Geese-Go-Away Month, and Wind Month. Likewise the Natchez on the lower Mississippi River had Deer Month, Little-Corn Month, and Bear Month.

Such specific natural cues must have come out of a long and close study of local fauna, flora, and other natural surroundings and events, information learned and then passed down over the generations both informally from parents to children and more formally as lists of months and easily remembered poems and calendar stories told and retold. Eventually these oral versions of time-reckoning cues were carved in stone and recorded on scrolls and parchments.

For instance, the Greek poet Hesiod (fl. ca. 800 B.C.) some 2,800 years ago took his local oral calendar used since ancient times in Peloponnesia and wrote it down in a long poem called *Works and Days*. A practical guide for organizing time, *Works and Days* is also a moral mandate to follow ancient rules of time and duty, which is not the first or last time that a calendar was used to codify standards of conduct. Hesiod wrote his poem at a time when Greece was becoming a maritime power in the eastern Mediterranean and many young men were turning away from farming and the discipline of the land to embrace commerce, war, and politics. The first part of the poem, "Works," is addressed to Hesiod's younger brother Perses, apparently one of those young men uninterested in the traditional life. Hesiod believed his wayward sibling needed some stern guidance from his older brother. But the spine of the story is about time:

> *Keep all these warnings I give you, as the year is completed*
> *and the days become equal with the nights again, when once*
> *more the earth, mother of us all, bears yield in all variety.*

Hesiod here refers to the most basic natural clock available—day and night—which in their respective lengths over the course of a year offer a crude guide to the seasons. He also alludes to cues such as the arrival of snails in early spring:

> *But when House-on-Back, the snail crawls from the ground up*
> *the plants . . . it's no longer*
> *time for vine-digging;*
> *time rather to put an edge to your sickles,*
> *and rout out your helpers.*

And of course Hesiod's poem invokes that other fantastic clock in the nighttime sky, the stars, using the position of constellations to guide his "exceedingly foolish" brother:

> *Then, when Orion and Seirios are come to the middle of the sky,*
> *and the rosy-fingered Dawn confronts Arcturus,*
> *then, Perses, cut off all your grapes, and bring them home with you.*
> *Show your grapes to the sun for ten days and for ten nights,*
> *cover them with shade for five, and on the sixth day*
> *press out the gifts of bountiful Dionysos into jars.*

But the most important time guide for Hesiod is the moon. This becomes evident in the second part of the poem, "Days," which treats time as a mystical force and the calendar as a cycle of lucky and unlucky days, of omens and sacred ceremonies. "Days" is structured around the 29 or 30 days of each Greek month and the phases running from one new moon to the next. It lists holy days, unlucky days, days that are ill-omened for the birth of girls, and the best days to geld bulls and sheep. "Avoid the thirteenth of the waxing month," writes Hesiod in a typical passage in "Days," "for the commencing of sowing / But it is a good day for planting plants."

The moon also gave Hesiod and the Greeks their year, which they based on 12 lunar months averaging close to 29½ days, equal to some

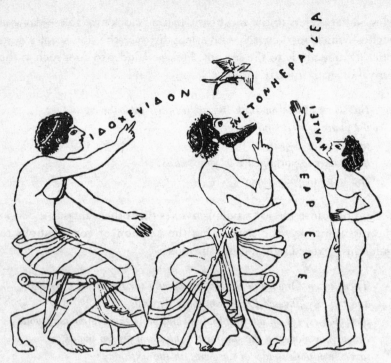

Early Greeks Greet Spring
"Look at the Swallow," says the man on the left, "So it is, by Herakles."
"There it goes!" "Spring is here!" From a vase, Fifth Century, B.C.

354 days. Nor were they alone. From ancient Sumer and China to the now-vanished Anasazi in Arizona, the moon became paramount, with variations on this same 12-month, 354-day year popping up everywhere as the Stone Age melded into the Neolithic age and people began to discover how to build cities, irrigate fields, set up governments, and organize armies to fight wars.

But alas, Luna was a mere temptress where time was concerned, drawing calendar makers down a false path—the first of many in humanity's struggle to create an accurate calendar. For dependence on the moon caused a serious error—much worse than the flaw that outraged Roger Bacon several millennia later. All he had to worry about were the 11 minutes or so that his calendar was running fast. The ancient Greeks and others who threw their lot in with the moon found themselves with calendars running almost 11 *days* fast, a misalignment that within a few years flings a calendar into disarray against the seasons, flip-flopping the summer and winter solstices in just 16 years. This situation is unacceptable to anyone using such a calendar as a guide to planting and harvesting, or for knowing the proper seasons for sailing, building houses, and worshiping gods.

The problem comes in the time it takes for the moon to pass through its phases as it orbits the earth, which is not a tidy number for dividing into a year of approximately 365¼ days.* In fact a true lunar month is a cumbersome 29.5306 days long as measured by modern instruments, equal to a precise 12-month lunar year of 354.3672 days. Stack that up against the true solar year of some 365.242199 days and one can appreciate the intense frustrations of astronomers over the centuries trying to link up the sun and moon.

As ancient cultures matured, frustration with the lunar drift stimulated their scientists and priests to ponder a solution—a line of inquiry that continues to this day as we try to fine-tune our days, weeks, and months to fit into the true solar year. But for the ancients, lacking modern tools and concepts, even approximating this year using the moon proved immensely difficult. A number of solutions were tried, but all failed.

The ancient Babylonians, for one, stuck with the moon despite their highly advanced knowledge of astronomy. But their infatuation was tempered by a compromise with the sun in what is now called a "lunisolar" year. Sometime around 432 B.C. Babylonian mathematicians figured out that seven years of thirteen lunar months followed by twelve years of twelve lunar months would equal almost exactly nineteen solar years. This later became known as the Metonic cycle, after the Greek astronomer Meton (c. fifth century B.C.). It works by

---

*This is called a *synodic month*.

inserting or "intercalating" extra months into the standard 12-month lunar year. But even this 19-year system is not completely exact, running several hours fast. It also proved unwieldy and impractical for everyday use, since few people were willing to keep track of such a complicated system over so long a time.

Other ancient cultures unwilling to give up the moon devised other systems of intercalations. The Greeks added an extra 90 days every eight years to compensate for their standard lunar calendar of 354 days, though the months were not always added on schedule and were often inserted haphazardly. The Jewish calendar intercalates a month every three years, inserted just before the month of Nisan, though this system still leads to a gradual drift that requires a second extra month to be added now and then by Jewish elders. According to legend, Chinese mathematicians, under orders from the Emperor Yao (c. twenty-fourth century B.C.), began experimenting with a calendar in 2357 B.C. that eventually became Metonic, adding seven months to the lunar calendar every 19 years.

The Sumerians by the twenty-first century B.C. had developed a slightly different system founded on a calendar year of 360 days. This came from rounding off the lunar month to 30 days, which fit neatly into the Sumerians' mathematic and astronomic system. This system is based on the numbers 6 and 60, which equal 360 when multiplied— the number we still use to divide the sky and every circular plane. No one knows why the Sumerians and later the Babylonians chose these numbers, though four thousand years later they remain the numeric basis for everything from determining one's position at sea to the location in the sky of a distant galaxy vis-à-vis the earth.

The Babylonians inherited and refined the older Sumerian numerology to divide the day up into 24 hours, which is divisible by six and also divides evenly into 360. Again, the reason for using 24 has been obscured by time, though it's likely that it had something to do with the zodiac, which the astrology-crazed Babylonians used with great fervor to guide their lives. Possibly they divided the day and then the night into 12 hours each to correspond to the signs of the zodiac, adding them together to reach the 24-hour day we still follow.

In the fifth century B.C. the Greek historian Herodotus told a story that points up the complications with these less-than-perfect luni-solar calendars. In *The Histories* Herodotus tells how the Greek lawgiver Solon once answered a question put to him by the rich and haughty Croesus of Sardis: Who was the happiest man he had ever seen? In answering, Solon refused to name Croesus, explaining that fate could still render him unhappy. He used the Greek calendar to emphasize his point. "Take seventy years as the span of a man's life," says Solon. "Those seventy years contain 25,200 days, without counting intercalary months. Add a month every other year, to make the seasons come round with proper regularity, and you will have thirty-five additional months, which will make 1,050 additional days. Thus the total of days for your seventy years is 26,250, and not a single one of them is like the next in what it brings. You can see from that, Croesus, what a chancy thing life is. You are very rich, and you rule a numerous people; but the question you asked me I will not answer, until I know that you have died happily."

Egypt was the first ancient civilization to correct the error of the moon and embrace the sun. Remarkably, they did it quite early—almost six thousand years ago, when people living along the Nile figured out the solar year was very close to 365 days. This led to a calendar with 12 months of 30 days each and an additional 5 days that Egyptian mythology says were added to the year by the god Thoth. These became the birthdays of Osiris, Isis, Horus, Nephthys, and Set.

How these Neolithic Egyptians figured out so close an approximation to the true year remains a mystery. Egyptian science was advanced very early, but they were never renowned for their astronomy, like the Babylonians, or for a keen interest in mathematics, like the Greeks.

The most plausible explanation is the Nile. Herodotus called Egypt "the gift of the Nile," and anyone who has visited understands instantly the division between the green along the river and the brown of the desert, between life and death. The Nile was responsible for the crops, the commerce, and the continuity of Egypt. The ancient Egyptians called it simply "the sea." Flooding from late June till late October, the Nile each year brought down rich silt for crops to be grown from October to February, and harvested from February until the end of June. These were the three seasons of life in Egypt: flooding, growth, and harvest. The regularity of this cycle and the availability of the great river as a natural timepiece provided an easy and dramatic alternative to the moon.

Northeast Africa was not always dependent on the Nile. Until the final retreat of the glaciers 10,000 years ago the Sahara was covered not with sand but with savanna. Then 7,000 or 8,000 years ago the savanna died as the earth warmed and the people of the northeastern Sahara were forced into the valley of the Nile. There they abandoned their Paleolithic life of hunting and gathering and adapted to the cycles of the river. This provided a deep-set regularity to the Egyptian culture, which began farming and building settlements by about 7000 B.C. Three millennia later Egyptians established what may be the first known date in human history, which chronographers have calculated to be as early as 4241 B.C. A thousand years later the kingdoms of the Nile united politically, launching a complex and homogeneous civilization with a central authority and religion that persisted with few breaks for three thousand years, until the death of Cleopatra, all the while depending on the rhythms of the great river.

The Nile is a gift of life; but it also is an enormous clock and calendar stretching over four thousand miles, the second-longest river in the world. Fed by rainfall and melting snow in the Ethiopian highlands and to a lesser extent by watersheds as far south as Uganda, the Nile floods with a predictability that Egyptians understood long before stone temples and pyramids began to rise on the river's shoreline—or before anyone thought about a formal calendar. All an early Egyptian farmer needed to do was plant a tall reed in the mud along the river bank, cut a notch to measure the high point of the floods, and then count the days until the next high-water mark, which would occur almost exactly one year later. This simple device, called a nilometer, was then the most accurate calendar in the world, based on the seasons as regulated by the earth's orbit and the tilt of its axis rather than on the phases of the moon.

Egyptian astronomers supplemented the nilometer with another discovery that made their solar year even more accurate: that Sirius, the Dog Star and the brightest star in the sky, ascends in the dawn sky once a year in a direct line with the rising sun. Sirius's appearance happened to coincide with the Nile's annual flood; it also became the first day of the month of Thoth, the Egyptian New Year's Day, commemorated annually with elaborate ceremonies that began when Sirius appeared on top of obelisks precisely aligned with observation points on the ground below. By timing Sirius's appearance exactly from year to year, Egyptian astronomers eventually realized that the solar year was one fourth of a day longer than 365 days. Egyptians also used pyramids to measure shadows to determine the coming of the equinoxes.

Adding a quarter of a day to the Egyptian year was a revolutionary discovery. It brought the Egyptian year within 11 minutes and 24 seconds (give or take a few seconds) of the true solar year at least two thousand years before Julius Caesar embraced the 365¼ day calendar for Rome, and over three millennia before Roger Bacon's appeal to Pope Clement. Still, in a move Bacon would have ruefully understood, the priests who controlled Egypt's calendar refused to alter their year to make the correction from 365 to 365¼ days. As orthodox and unbending as the Catholic Church in Bacon's era, the white-kilted Egyptian priests with their shaved heads and painted faces considered their calendar too sacred to alter, leaving it to shift by six hours (a quarter of a day) each year. This launched the Egyptian calendar on a slow drift across the seasons in a cycle that repeated itself every 1,460 years. Called the Sothic cycle, this flaw was not corrected until the Ptolemaic era in Egypt. In 238 B.C. Ptolemy III★ ordered a leap-year system by adding an extra day every four years. But even then the priests resisted the edict until 30 B.C., when Rome conquered Egypt and Augustus forced the people of the Nile to add the extra quarter of a day to their calendar to bring it into line with the Julian calendar. This stabilized the Egyptian calendar so that the first of Thoth always fell on August 29.

Egyptians were not alone in their early turning to the sun. Far beyond the great Nile valley and even the Mediterranean, on the distant edge of the Eurasian continent, a little-understood people also figured out a close approximation of the solar year a few centuries after the Egyptians. We know this only because they left behind what appears to be an enormous calendar constructed out of immense slabs of bluestone, standing upright to form megaliths, some of them topped by lintels called *henges*. Standing on the barren Salisbury plain, this structure, Stonehenge, was used for over two thousand years by ancient Britons, who aligned the stones so that at the precise moment of the summer solstice a ray of sun shines down the main avenue and into its center. But what was this for? Is Stonehenge truly an enormous calendar? Or is it an observatory, a fortress, a temple, a Bronze Age place of assembly—or all of the above?

No one knows for sure, though the layout leaves no doubt that the people who built it were astronomically sophisticated enough to build a device to accurately measure the solar year. Further evidence comes from stones erected in patterns around Stonehenge that align with the

---

★The royal dynasty of the Ptolemies should not be confused with the second century Alexandrian astronomer Claudius Ptolemy.

sun at both solstices and at the equinoxes, and with the moon as it runs through its orbit around the earth. This giant calendar would have allowed an ancient Briton to anticipate astronomic cycles and events as accurately as the Egyptians watching Sirius—or, for that matter, a modern astronomer using solar and star charts. Some have claimed that Stonehenge can also foretell eclipses of the moon, which occur regularly after those months when the full moon rises precisely down the main avenue.

The other ancient culture that invented sun time early on, the Maya, was far more isolated than the people of Wessex. Raising great cities filled with temples and palaces deep in the interior of Central America, the Maya also invented a calendric system so accurate that when the Spaniards conquered them in the sixteenth century, the Julian calendar the conquistadors brought with them was less precise.

The Maya developed three calendar systems. The first was 365 days long, with 18 months of 20 days, to which they added 5 more days. This was called the *haab*. As for the Egyptians, these five extra days were considered special, though the Maya believed them to be unlucky and shunned all activity as they anxiously waited for them to pass. Apparently the Maya knew that the year was really closer to 365¼ days but ignored it in this calendar, which drifted, like the Egyptian version, about six hours a year. Concurrently with this 365-day calendar the Maya used a 260-day cycle called a *tzolkin,* or "sacred round," which served a similar purpose as Hesiod's "Days," listing omens and associations for each day to guide the Maya and other Mesoamericans in planting, waging war, and offering sacrifices to the gods. The 260-day cycle was developed early in the first millennium B.C. by the Zapotecs of Mexico, for reasons that remain obscure. Common to all sophisticated Mesoamerican peoples by the time of the Maya, who first appear in about 1000 B.C., the *tzolkin* was joined to the 365-day calendar in a complex cycle of 52 years called the Calendar Round. This is the time it takes for both calendars to start again on the same day. Spanish conquistadors in the sixteenth century reported that the end of a 52-year cycle was commemorated by all advanced cultures in the region. It was universally greeted with great despondency, the people fearing that the sun might not return.

The third Maya calendar was the Long Count, used to calculate long periods of time. It was based on 360-day unit called a *tun* and a number system based on 20 (Mesoamericans counted with their fingers *and* their toes). The Long Count cycles are as follows:

20 *kins*	=	1 *uinal*	=	20 days
18 *uinals*	=	1 *tun*	=	360 days
20 *tuns*	=	1 *katun*	=	7,200 days
20 *katuns*	=	1 *baktun*	=	144,000 days

The Mayas multiplied the baktun by 13 to get what they termed a Great Cycle, equal to 5,130 years.

At the same end of a great cycle, the Maya, Aztecs, and other Mesoamerican peoples believed all things would cease to exist and an entirely new world would be ushered in to start the next great cycle. The current great cycle probably began in 3114 B.C. and will end on December 23, 2012.

Presumably the Maya discovered the true solar year using natural cues and careful astronomical observations, though exactly how they did it remains a puzzle. Until recently scholars believed their motivation was a literal worship of time, though new interpretations since the breaking of the Maya language code reveal that the Maya actually used their calendars to legitimize the acts of kings and other key events by recording with great accuracy the day, hour, and even minute when they occurred. This is shown in countless hieroglyphics, steles, and paintings depicting the exact date when specific kings and queens waged battle, ceremonially mutilated themselves, married, and performed important sacrifices.

Maya signs for the months in the 365-day count

Maya and other Mesoamerican gods also seem to have demanded that their priests perform ceremonies precisely on time. Nowhere was this taken more seriously—and to such a bizarre extreme—than among the Aztecs. Obsessed with the belief that they must keep time on its proper course, the Aztecs offered a numbing progression of human sacrifices to appease their sun god, Tonatiuh, to assure that he would rise each day and cross the sky.

The Aztecs believed that the sun required for "fuel" rivers of blood from victims who ranged from priests and criminals to the deformed, though most were prisoners captured in warfare. If Spanish chroniclers can be believed, the Aztecs sacrificed 20,000 to 50,000 people a year in their capital, Tenochtitlán, with each month requiring a prescribed tally of victims: male and female, child and adult. For instance, in the months when the rains were supposed to come, children were drowned or walled up in caves. The more they wept and cried, the better the omen for rain. Others were flayed to help crops grow and burned to death during harvest time. To feed the need for such huge numbers of victims, the Aztecs arranged a peculiar agreement with their neighbors to fight regular ceremonial battles not for conquest, but to allow each side to capture large quantities of sacrificial victims. Apparently most of the victims seized in what was called the War of Flowers considered sacrifice an honor or an unquestionable act of fate. Most were anesthetized first with narcotic plants, though all were left conscious enough to scream and exhibit pain, which was part of this bloodiest of time rituals.

Despite the remarkable achievements in time reckoning by Mesoamericans and the people of Wessex, out of all those who early embraced the sun, it is the Egyptians who lie in the direct path of our story. It is their affair with Sol that brought us our calendar, making the solar year victorious over the moon first along the Nile and then in Europe, and much later around the world. But this triumph of the Egyptian year was hardly inevitable. Nor was it even likely given the circumstances that led to the fusing of the ancient solar calendar of the Nile with a brash, upstart empire ruled by a people living on another river, the Tiber, and led by a conqueror whose adoption of a new calendar had more to do with his love for a legendary woman than with a passion for accurately measuring time.

3
Caesar Embraces the Sun

Caesar . . . reorganized the Calendar which the College of Priests had allowed to fall into such disorder, by inserting days or months as it suited them, that the harvest and vintage festivals no longer corresponded with the appropriate seasons.

—SUETONIUS, A.D. 96

As night fell, a small ship slipped under the sea chain defending Alexandria's harbor, raised by guards bribed to let it pass. The boat on this balmy October evening in 48 B.C. stole quietly through the black waters, past quays and warehouses full of grain and treasure. Skirting the fleets of Egyptian and Roman warships, the boat carried a cargo that would not only transform two great empires, but lead to a revolution in measuring time that is directly responsible for calendars hanging on walls from present-day St. Louis to Singapore.

After the boat landed unnoticed on a stone wharf, a Sicilian named Apollodorus leapt ashore, carefully lifting onto his back a rolled-up coverlet tied at each end. Apollodorus carried his load past Roman sentinels, explaining by the light of torches that he bore a gift for the recently arrived Julius Caesar, dictator of Rome. Led to the general's apartment in Alexandria's royal palace, Apollodorus greeted Caesar by unfurling the coverlet, which concealed a woman.

She can hardly have appeared dignified emerging from a bedroll. Yet as Cleopatra rose in front of the astonished Caesar, she managed to impress him profoundly with her majesty and sexual allure—and also with the pathos of a woman who desperately needed help from the most powerful man in the Western world.

Cleopatra's trouble had begun a few months earlier when her teen-aged brother and coruler, Ptolemy XIII, staged a palace coup with his advisors and forced her to flee the city. Escaping to Syria, she had recently returned to Egypt at the head of a small army, determined to wrest back her throne—a cause she hoped to convince the newly arrived Caesar to embrace.

Poets and romantics tell us Caesar was smitten from the moment he saw Cleopatra. She was twenty-two years old and a queen since her father, Ptolemy XII, had died three years earlier, leaving her and her then ten-year-old brother to jointly rule in the Egyptian fashion. Cunning, brilliant, and erotic, Cleopatra spoke several languages, was highly educated in science and literature, and was possessed of an insatiable ambition that amused and captivated the master of the Roman world. The Roman poet Lucan (A.D. 39–65) says the general and the queen made love that very night.

Caesar was fifty-two years old at the time. "Tall, fair, and well built," according to the Roman historian Suetonius, but also balding and epileptic, he was on the verge of becoming the undisputed dictator of an empire that had just conquered virtually the entire Mediterranean world and parts beyond. Caesar himself had seized Gaul in a series of masterly victories ten years earlier. Since then he had been locked in a wrenching civil war against another brilliant general and conqueror, Gnaeus Pompeius Magnus—Pompey for short. Caesar had just arrived in Egypt in hot pursuit of Pompey, who fled there after a crushing defeat by Caesar in the Battle of Pharsalus in central Greece. Arriving three days after Pompey, Caesar had been welcomed off the coast of Alexandria with a grisly gift from the boy-king Ptolemy and his advisors: General Pompey's embalmed head wrapped in Egyptian linen. A soldier hired by Ptolemy's court had stabbed the great general in the back as he stepped off his boat. Caesar reportedly wept at the specter of this great Roman being assassinated by foreigners. But his sorrow was tempered with relief if not a carefully concealed elation, for the empire was now his.

With Pompey dead, Caesar should have left for Rome to consolidate his victory. Instead he stayed to settle the conflict in Egypt, a country still nominally independent but in thrall to Rome, and to be with Cleopatra. The latest in a never-ending string of mistresses— Caesar's troops sang of his conquests in battle *and* in bed when they celebrated his triumphs—Cleopatra impacted both his libido and his politics. "Overcome by the charm of her society," writes the Roman biographer Plutarch, he forced the boy-king Ptolemy within days of Cleopatra's dramatic entrance to reconcile with his sister, ordering "that she should rule as his colleague in the kingdom." Cleopatra then promptly threw a party to celebrate—which is where Caesar first heard about the Egyptians' solar calendar, according to Lucan.

This seems an unlikely venue for an event that would literally re-order time for millions of people. Indeed, Lucan tells us Cleopatra hardly had calendar making on her mind the night of the soirée. Dressed in heavy strands of pearls, "her white breasts . . . revealed by the fabric of Sidon," and her hair wrapped in wreaths of roses, she seemed far more intent on dazzling her lover with the riches and exotica of Egypt: "birds and beasts" served on gold platters, crystal ewers filled with Nile water for their hands, and "wine . . . poured into great jeweled goblets."

Still, Eros and fine food were not all that the young queen and her court offered to this uncommonly curious Roman conqueror. "When sated," says Lucan, Caesar began discoursing with a scholar attached to the royal court, an elderly wise man named Acoreus, "who lay, dressed in his linen robe, upon the highest seat." Caesar asked questions about the source of the Nile, the history of Egypt—and about the country's calendar. It was during this conversation that Caesar heard about Egypt's reliance on the sun for its year—measured by the annual rise of Sirius in the eastern sky and by the flooding of the Nile, which, the Alexandrian sage said, "does not arouse its water before the shining of the Dog-star."

No other ancient source that I am aware of describes this scene or mentions the sagacious Acoreus, although something like this un-doubtedly happened to inform Caesar about the Egyptian system for measuring time. Later he would take this new knowledge back to Rome, though for the moment he seemed in no rush to leave.

Caesar's liaison with Cleopatra was also an infatuation with Egypt itself. Already very ancient even in Caesar's day, this was a country of fantastic wealth and mystery—and, during the final years of the Ptolemaic dynasty, of a decadence and sensuality very foreign to a Roman raised in the austerity of the republic. But Alexandria was also a feast for the mind, a city that even in its decline as a regional power remained one of history's premier centers of learning and sophistication. For three centuries it had attracted the greatest minds of the far-flung Hellenistic world, who created a milieu of intellectualism that fostered a breathtaking progression of discoveries—including groundbreaking work on time and the calendar.

Founded by Alexander the Great when he conquered Egypt in 332 B.C., the city was seized after Alexander's death by Ptolemy, one of his key generals. Declaring himself king of Egypt in 305 B.C., Ptolemy I lavished the Nile valley's wealth on his new capital, creating a haven for scholars who came from as far away as India, which was briefly connected to the Hellenistic world after Alexander's conquests. The city quickly expanded to at least 150,000 people, one of the largest in the ancient world, as Ptolemy and his dynasty filled the city with magnificent palaces, temples, gymnasiums, museums, and amphitheaters. Sometime around 307 B.C. the Athenian writer and statesman Demetrius of Phaleron inspired Ptolemy I to lay the foundations for the great library of Alexandria, which eventually housed hundreds of thousands of papyrus scrolls, including Aristotle's personal library. A generation later Ptolemy II (308–246 B.C.) built the famed Pharos lighthouse, one of the Seven Wonders of the Ancient World, towering four hundred feet and emitting a blazing fire signal that could be seen for miles offshore.

Luminaries during Alexandria's golden age included Apollonius of Rhodes, author of the *Argonautica*, about Jason's quest for the golden fleece; the anatomist Herophilus of Chalcedon, who performed one of the first systematic autopsies; and Euclid and Archimedes, whose ideas form the core of Western mathematics. But perhaps the greatest achievements in this city on the western edge of the Nile delta, hard by the Libyan desert, were a long line of discoveries in astronomy, some of which became the basis for the new calendar born of Caesar's tryst with Cleopatra.

The stargazers of Alexandria started with the patrimony left them by earlier Greek astronomers and mathematicians. Since at least the sixth century B.C., they had been looking up in the sky and postulating about what they saw. The earliest of these postulated that the sun is one foot wide and is renewed afresh each day, and that the earth either floats on water or is supported on air. But they also realized that "moonshine" is really reflected sunlight, that the moon is closer to the earth than to the sun, and that eclipses are caused by the shadow of the earth and other celestial bodies.

These speculations gave way to more solid science with Pythagoras (sixth century B.C.), who developed some of the early geometry and mathematics used by later astronomers to analyze the respective positions of the sun, moon, earth, and stars. Then came the Athenian astronomer Meton, discoverer of the Metonic cycle in 432 B.C. At roughly the same time the astronomer Euctemon estimated the length of the seasons, though he got them wrong. A century later Callippus of Cyzicus calculated the correct lengths to round figures—90 days for summer, 90 for autumn, 92 for winter, and 93 for spring. Also working in the fourth century B.C., the astronomer Eudoxus of Cnidus devised a mathematical theory involving spheres that he used to try to explain the motions of the planets and the moon, and what appeared to be the motion of the sun in an earth-centered universe. Aristotle (384–322 B.C.) also weighed in, working in the years immediately leading up to the founding of Alexandria. His writing in astronomy expands on Eudoxus's theory of the planetary spheres by suggesting that the stars, planets, and sun literally are encased in invisible spheres that orbit the earth in a series of concentric circles.

One of greatest of the early astronomers in Alexandria itself was Aristarchus (fl. c. 270 B.C.), who constructed a modified sundial called a *skaphe*—a spherical bowl with a needle standing up in the center like a miniature obelisk to cast shadows against lines marked off on the bowl's surface. Using this device he could measure the height and direction of the sun. This allowed him to figure out that the sun shines light against a half moon, as seen on earth, at an angle of 87 degrees. From this he surmised that the sun is many times the size of the earth and must be very far away.

Aristarchus also deduced that the earth circles the sun, an astronomic theory that ran counter to the accepted orthodoxy that the sun orbited a stationary earth. He argued that the sun *seems* to move across the sky because the earth spins on its axis. But lacking a telescope and accurate star charts, Aristarchus could not prove something considered ludicrous by an earth-centered world, one that would remain convinced the sun was subservient to our little planet for another eighteen centuries, until the age of Copernicus and Galileo.

A generation after Aristarchus, the Alexandria-based mathematician, philosopher, geographer, and astronomer Eratosthenes (276–194 B.C.) deduced within a tenth of a degree the tilting of the earth's axis of rotation, which causes the seasons. He also measured the circumference of the earth to within 250 miles of the true value. A few years later Ctesibius of Alexandria constructed an elaborate water clock using floats, a chain winch, cog shaft, dial, and a sundial system that linked the path of the sun astronomically and geometrically with levels of its shadow.

In about 130 B.C. the astronomer Hipparchus (fl. 146–127 B.C.) discovered the precession of the equinoxes, a slow shift westward of the equinoctial points against the stars, something Isaac Newton much later determined was caused by the very subtle gravitational tug of the moon and sun on the earth. Hipparchus published a celestial catalogue, since lost, that described hundreds of stars and provided calculations about distances among them. He also confirmed the accuracy of the Egyptian year by studying several years' worth of solstices to come up with a reasonably close approximation of the true solar year: 365 days, 5 hours, and 55 minutes, some six minutes too long.

But none of these stargazers were as influential as Alexandria's last great astronomer, Claudius Ptolemy. A Greek and a citizen of Rome who flourished some two centuries after Caesar's sojourn in Egypt, Ptolemy compiled during the second century A.D. a massive encyclopedia on astronomy and geography that became, with Euclid's *Elements* on mathematics, a widely revered if not always understood textbook in the Middle Ages. Ptolemy's calculations about the length of the month and year; the motions of the sun, moon, and stars; eclipses; and the precession of the equinoxes became the benchmarks used by every time reckoner who followed him for over a thousand years: Bede, Roger Bacon, and the chief architects of the calendar reform in 1582, Christopher Clavius and Aloysius Lilius. Ptolemy's value for the length of the solar year, which he borrowed from Hipparchus, happened to be wrong by several minutes. Yet it is worth noting that Ptolemy and the Alexandrians knew Caesar's year of 365¼ days was in error centuries before Roger Bacon—and some 1,400 years before Pope Gregory finally fixed it.

On the night of Cleopatra's feast Caesar may have gotten an earful about Egypt's calendar, but as it turned out he almost missed his chance to use it. That very night he narrowly avoided being killed in an attempted palace coup. Only the intervention of Caesar's barber, a busybody who overheard the plotters, saved him. As it was, Caesar had barely enough time to protect himself and to muster his troops. After fierce fighting inside the palace, the general and his men managed to secure the royal compound, though this left them under siege by the boy-king's army and a mob of anti-Roman Alexandrians. The

Romans retained access to their small fleet, moored to the palace docks, but were blockaded from leaving the main harbor by Egyptian warships.

Foolishly, Caesar had come to Alexandria with only two depleted legions from the battle at Pharsalus. No more than 3,200 men and 34 ships were pitted against an Egyptian army numbering at least 22,000 men supported by a large Alexandrian navy. Fortifying the palace and securing the royal harbor, Caesar dispatched messengers to fetch reinforcements from his legions in Syria and Greece. He then launched a series of sorties to reinforce his position, at one point setting fire to part of the Alexandrian fleet. Tragically, these flames spread to the shore, destroying several buildings in the lavish Brushium district west of the palace, including buildings that housed part of the great library's priceless collection. In another skirmish, fought over a causeway connecting the island of Pharos to the city, Caesar's position was overrun, forcing him to swim for his life to a Roman skiff, pelted all the way by Egyptians who could easily single him out in his imperial purple toga.

Caesar ultimately prevailed, however, when a large relief force of legionnaires arrived some five months later. With these he crushed his enemy and restored his lover to her throne.

Caesar was now free to return to Rome, but delayed again, this time to celebrate his victory with a two-month journey with his mistress down the Nile. Luxuriating on an immense barge filled with banquet halls and apartments fitted out with cedar, cypress, ivory, and gold, the general and the queen feasted, relaxed, and made love, producing in due time a son that Caesar would later recognize as his own, calling him Caesarion. Hoping to float all the way to Ethiopia to discover the source of the Nile, Caesar during this trip undoubtedly continued his discourse with the sages of Egypt. These may have included a court astronomer named Sosigenes, who wrote several books about the stars, all of them now lost. But unlike those great stargazers whose works *have* been preserved, Sosigenes at some point during Caesar's time in Egypt passed on something far more lasting than suppositions about the placement of stars and the distance of the sun and moon: a breathtakingly simple idea for reforming the Roman calendar.

In June of 47 B.C., Julius Caesar finally departed Egypt. As a parting gift he left the pregnant Cleopatra three Roman legions to protect her, but also to guard the interests of Rome against a woman Caesar clearly understood was as ruthless as he in her ambitions. Desperately needed in Rome to sort out the aftermath of the civil war, Caesar first launched two lightning-quick wars against an upstart king in Syria and against the remnants of Pompey's army, which had fled to the north coast of Africa. He then returned to Rome, where the Senate named him dictator for ten more years, commissioned a bronze statue of him to be erected in the Forum, and ordered a celebration of forty days for his victories in Gaul, Egypt, Syria, and Africa. This triumph became a legendary orgy of festivals, games, and debauches that included the slaughter of four hundred lions in the Circus, and mock battles on land and sea in which hundreds of war captives and criminals died. For days at a time Caesar's soldiers marched in parades leading into the Forum, carrying more than 20,000 pounds of captured treasure and leading in countless prisoners weighed down by chains. These included the young princess Arsinoë, a sister of Cleopatra who had sided with her enemies.

Caesar's supporters reveled in their triumph, though many Romans, raised in a republic that had for centuries despised the idea of a king, found the celebrations grossly ostentatious and an unsettling display of arrogance and personal power. The Roman historian Dio reports that people recoiled against the bloodshed and the "countless sums" lavished on the shows. People also complained about the treatment of high-born prisoners, including Arsinoë. Demeaned in her chains, she "aroused very great pity," to the point that Caesar released her rather than face the wrath of the populace. Not even a lavish gift of gold, grain, and oil to every free person in Rome assuaged a general anxiety about what Caesar would do next. Already his enemies were talking darkly of a man whose success and virtually limitless power were turning him into a monster.

The fact that Caesar governed mostly with energy and resolve after his infamous fete made his enemies revile him even more, since an able dictator set back the cause of those who longed for a return of the republic far more than if Caesar had been inept. He plunged into a dizzying series of projects ranging from a flurry of new temples and a planned canal across the Isthmus of Corinth to hundreds of new laws and reforms. He dissolved the corrupt guilds in the city; limited the terms of office for senior elected officials; forgave a quarter of the debts owed by all Romans, to stimulate the economy; awarded prizes to large families to increase the population, depleted by the war; and reduced the expensive subsidies of grain to the city's paupers. He also consolidated power by naming his own men to key offices and by co-opting control of the Senate.

But none of the measures taken by Caesar during his first months back in Rome was more dramatic than the one he decreed sometime in the first half of 46 B.C.; the reordering of the Roman calendar. More than a simple adjustment in the way days were counted, this reform was a potent symbol not only of Julius Caesar's newfound authority but also of an empire that believed it had the power to reorder time—not only for its own people but for subjects living in far-flung locales, from the English Channel to what is now Iraq. Fortunately for the millions of people who would have to use his calendar, Caesar's hubris coincided with the pragmatism of a veteran general and statesman who based his new calendar on science, not vanity or religious dogma. In any case, Rome's old lunar calendar was in desperate need of reform, running in Caesar's day several months fast against the solar year.

Like many other ancient cultures, the Romans centuries earlier had developed a system based on a 12-month lunar year, plus occasional days and months intercalated by priests to keep the calendar year more or less in line with the seasons. But over the centuries the calendar had drifted back and forth because the priests either neglected to insert extra months or because they intentionally manipulated the calendar for political reasons. For instance, the highly politicized college of priests sometimes increased the length of the year to keep consuls and senators they favored in office longer, or decreased the year to shorten rivals' terms. The college also misused their calendar to increase or decrease taxes and rents, sometimes for their personal financial advantage.

By legend, the Roman calendar—our calendar—was created by the mythic first king of Rome, Romulus, when he founded the city in 735 B.C.—year 1 in the Roman calendar, known as *ab urbe condita* (A.U.C.), "from the founding of the city." But unlike most moon-based calendars, Romulus for some unknown reason concocted a year composed of only 10 months, not 12, for a year that totaled 304 days. The ancient Roman poet Ovid (43 B.C.–A.D. 17), who wrote poems about love and about the calendar, submits that the erring warrior-king "was better versed in swords than stars," and may have been trying to emulate "the time that suffices for a child to come forth from its mother's womb"—a gestation period roughly corresponding to 304 days. Another reason may have been the Roman reverence for the number 10, says Ovid, "because that is the number of the fingers by which we are wont to count." Romulus repeatedly used the number 10 in organizing his new kingdom, dividing both the 100 senators and his military units of spearmen, infantry, and javelin throwers into groups of 10. Latin numerals themselves—I, II, III, IV, V, VI, VII, VIII, IX, X—are probably symbols meant to represent fingers counting up to 10, with the V perhaps equating to an upraised thumb and index finger and the X to an upraised palm.

Romulus's infatuation with ten extended to naming his months. In one of the more unimaginative bursts of calendar making ever, this ancient king started out attaching descriptive names to Roman months, then seems to have run out of ideas. The first four months he named Martius for the god of war; Aprilis, which probably refers to raising hogs; Maius, for a local Italian goddess; and Junius, for the queen of the Latin gods. Then he simply fell into counting the months, naming them the fifth, sixth, seventh, eighth, ninth, and tenth: in Latin Quintilis, Sextilis, September, October, November, and December. This mythic king's lack of attention explains for why the tenth, eleventh, and twelfth months of our modern calendar are still numbered in Latin as the eighth, ninth, and tenth months.

Romulus and his successors were equally unimaginative in their system of numbering days of the month. They divided up each month not into weeks, which were introduced in Europe much later, but into day markers that fell at the beginning of the month, on the fifth (or seventh) day, and in the middle. These three signal days were called *kalends* (the origin of our word *calendar*), *nones,* and *ides.* Most other days in the Roman calendar had no given name. Instead each was numbered in a confusing system according to how many days it fell *before* the kalends, nones, or ides. For instance, here is the Roman system for the first half of March:

Modern Date	Roman Date
March 1	Kalends Martius 1st
March 2	VI nones (5 days before nones)
March 3	V nones (4 days before nones)
March 4	IV nones (3 days before nones)
March 5	III nones (2 days before nones)
March 6	Pridie nones (day before nones)
March 7	Nones
March 8	VIII ides (7 days before ides)
March 9	VII ides (6 days before ides)
March 10	VI ides (5 days before ides)
March 11	V ides (4 days before ides)
March 12	IV ides (3 days before ides)
March 13	III ides (2 days before ides)

| March 14 | Pridie ides (day before ides) |
| March 15 | Ides |

Romans would refer to March 11, say, as "Five ides," which was as clear to any other Roman as someone today saying "March 11." Still, given the complexity of this system, it is amazing that it lasted some two thousand years, operating as the official dating system in Europe well into the Renaissance. As late as the seventeenth century, William Shakespeare could write his famous lines in *Julius Caesar,* "Beware the ides of March," and expect his audience to know what he meant.

Romulus's 304-day calendar was shorter-lived, being entirely unworkable for an agricultural people who needed a reasonably accurate calendar to guide them through the seasons. It was Romulus's successor, King Numa, who added two more months to the calendar around 700 B.C.—Januarius and Februarius. This brought the year to the standard lunar year of 354 days, to which Numa added another day because of a Roman superstition against even numbers.

This 355-day year was a considerable improvement over Romulus's calendar, though it did not take long for Roman farmers to figure out that it too was flawed and needed days and months intercalated to keep it in line with the seasons. The Romans attempted several schemes to make the correction, none of which worked very well. First they tried adding an extra month every two years. But they miscalculated its length and overshot their mark, coming up with a 366¼-day-long year on average. Realizing this calendar ran *slow* against the true year by a day and a quarter, the Romans adopted a version of the Greek calendar that inserted intercalary months every eight years. This brought their calendar roughly in line with a 365-day year, though the Greek system was so confusing that the priests frequently forgot to slip in the extra months at the proper interval or botched the job, causing calendric time to slip back and forth against the solar year.

There was also politics. From the beginning the Roman calendar, like most others, was a powerful political tool that governed religious holidays, festivals, market days, and a constantly changing schedule of days when it was *fas*, or legal, to conduct judicial and official business in the courts and government. (These *dies fasti*, or "right days," gave the Roman calendar its name: the *fasti*.) In the early days of Rome the calendar and the all-important list of *dies fas* were controlled first by the Roman kings and then in the early days of the republic by the aristocratic patrician class. For the first several centuries after Romulus the priests and aristocrats kept the calendar a secret they shared only among themselves, which gave them a tremendous advantage over the merchants and "plebs"—commoners—in conducting business and controlling the elaborate structure of religious auguries and sacrifices that governed much of Roman life.

This monopoly on official time ended in 304 B.C., when the plebs finally became so incensed that one of them, Cnaeus Flavius—the son of a freedman, who was later elected to several high offices—pilfered a copy of the codes that determined the calendar and posted it on a white tablet in the middle of the Roman Forum for all to see. After this the priests and patricians relented and issued the calendar as a public document—the first step in evolving the objective, secularized calendar that Caesar introduced two and a half centuries after Flavius's theft.

But Flavius did not entirely win the day, for the patricians retained an important prerogative as the class from which Rome's priests were drawn: control over when to insert intercalary months. It was this privilege that they abused so shamelessly for financial and political gain. Indeed, when Caesar returned home from Egypt and his other wars in 46 B.C., he found that the many years of misuse had left the calendar in a shambles. Caesar himself was in part to blame, since he had held the title of chief priest—*pontifex maximus*—for several years, and had inserted an intercalated month into the calendar only once since 52 B.C. This had left the Roman year to veer off the solar year by almost two full months. Perhaps this was an intentional manipulation by Caesar or his allies among the priests, or whether it was a simple oversight by a *pontifex maximus* distracted by civil war. What-ever the cause, it played havoc not only with farmers and sailors but also with a population becoming more dependent than ever on trade, commerce, law, and civil administration in a rapidly growing empire that desperately needed a standard system for measuring time.

To fix the calendar, says Plutarch, "Caesar called in the best philos-ophers and mathematicians of his time," including the Alexandrian astronomer Sosigenes, who seems to have come to Rome from Al-exandria to fine-tune the reforms he and Caesar had discussed in Egypt. The core of the reform was identical to the system ordered by Ptolemy III in 238 B.C.—a year equaling 365¼ days, with the fraction being taken care of by running a cycle of three 365-day years followed by a "leap" year of 366 days.

To bring the calendar back in alignment with the vernal equinox, which was supposed to occur by tradition on March 25, Caesar also ordered two extra intercalary months added to 46 B.C.—consisting of 33 and 34 days inserted between November and December. Com-bined with an intercalary month already installed in February, the entire year of 46 B.C. ended up stretching an extraordinary 445 days. Caesar called it the *"ultimus annus confusionis,"* "the last year of con-fusion." Everyone else called it simply "the Year of Confusion," re-ferring not just to the extended year but also to the heady whirlwind of change inaugurated by Caesar, who in effect was launching a vast new empire that already was profoundly reordering countless lives.

The extra days in 46 B.C. caused disruptions throughout the Roman world in everything from contracts to shipping schedules. The Roman historian Dio Cassius writes about a governor in Gaul who insisted that taxes be assessed for Caesar's two extra months. Cicero in Rome complained that his old political adversary Julius was not content with ruling the earth but also strove to rule the stars. Yet most Romans were relieved finally to have a stable and objective calendar—one based not on the whims of priests and kings, but on science.

To round out his calendar reforms, Caesar moved the first of the year from March to January, nearer to the winter solstice—an earlier calendar reform that had not always been adhered to. He then reorganized the lengths of the months to add in the ten days required to bring the year from 355 to 365 days, arranging them to create a calendar of 12 alternating 30- and 31-day months, with the exception of February, which under Caesar's system had 29 days in a normal year and 30 in a leap year. He left the old calendar largely intact in terms of festivals and holidays. He also retained the old system of numbering days according to kalends, nones, and ides, as well as the traditional names of the months, though later the Senate changed Quintilius to Julius (July) in his honor.

When the new day dawned on January 1, 45 B.C.—the kalends of Januarius, 709 A.U.C.—Romans awoke with a new calendar that was then among the most accurate in the world. Even so, it remained subject to errors and tinkering by priests and politicians. The first mistake was to come soon after Caesar's death in 44 B.C., when the college of pontiffs began counting leap years every three years instead of four. This quickly threw the calendar off again, though the error was easily fixed later by Emperor Augustus. Catching the mistake in 8 B.C., he ordered the next three leap years to be skipped, restoring the calendar to its proper time by the year A.D. 8. Since that year this calendar has never missed a leap year—with the exception of those century leap years eliminated by Pope Gregory XIII in his calendar reform of 1582. But Augustus and his handpicked Senate did not stop with this sensible and necessary calendar fix. They also tampered with the length of the months, with results far less satisfactory.

This Augustan "reform" began when the Senate decided to honor this emperor by renaming the month of Sextilis as Augustus. Part of the resolution passed by the Senate has been preserved:

> Whereas the Emperor Augustus Caesar, in the month of Sextilis, was first admitted to the consulate, and thrice entered the city in triumph, and in the same month the legions, from the Janiculum, placed themselves under his auspices, and in the same month Egypt was brought under the authority of the Roman people, and in the same month an end was put to the civil wars; and whereas for these reasons the said month is, and has been, most fortunate to this empire, it is hereby decreed by the senate that the said month shall be called Augustus.

This simple name change would have been fine. But either out of vanity or because his supporters demanded it, the Senate decided that Augustus's new month, with only 30 days, should not have fewer days than the month honoring Julius Caesar, with 31 days. So a day was snatched from February, leaving it with only 28 days—29 in a leap year. To avoid having three months in a row with 31 days, Augustus and his supporters switched the lengths of September, October, November, and December. This wrecked Caesar's convenient system of alternating 30- and 31-day months, leaving us with that annoying Old English ditty that seems to have originated in the sixteenth century, though it must have had precedents far more ancient:

> *Thirty days hath September,*
> *April, June and November,*
> *February has twenty-eight alone*
> *All the rest have thirty-one.*
> *Excepting leap year—that's the time*
> *When February's days are twenty-nine.*

Later Roman emperors would also try naming months after themselves. Nero, for instance, renamed April Neronius to commemorate his escape from an attempted assassination in that month during A.D.

65. Other month changes that failed to stick included substituting Claudius for May and Germanicus for June. When the Senate tried to change September to Tiberius, this taciturn emperor vetoed the measure, coyly asking: "What will you do when there are thirteen Caesars?" In the provinces local leaders and subject kings frequently changed their months to flatter powerful figures of the moment. In Cyprus the calendar once had months named for Augustus; his nephew Agrippa; his wife, Livia; his half-sister, Octavia; his stepsons, Nero and Drusus; and even Aeneas, the legendary founder of Rome from whom Julius Caesar, Augustus, and the entire Julian brood claimed to be descended.

The second error in Caesar's calendar was less easy to repair than the mixup over whether a leap year came every third or fourth year. This was the conundrum noticed by the Alexandrian astronomers Hipparchus and Ptolemy and later by Roger Bacon and others in the Middle Ages—that Caesar's year of 365 days and 6 hours ran slow. It's likely that Caesar's Egyptian advisor Sosigenes knew this too, though if he was familiar with Hipparchus's calculation, no one mentions it.

But even if Caesar was aware of a flaw amounting to a few minutes, it hardly would have upset him given the centuries that Romans had had to endure a calendric system that was frequently many days or months in error.* Indeed, after the initial grousing about the change, Caesar's calendar became an object of pride for educated Romans. Excavations throughout the former empire have yielded calendars carved in stone and painted onto walls, much as we hang our calendars today.

---

*Romans had not yet invented the idea of a "minute." They had a loose notion of an hour of the day; for astronomical purposes they divided up the day into simple fractions.

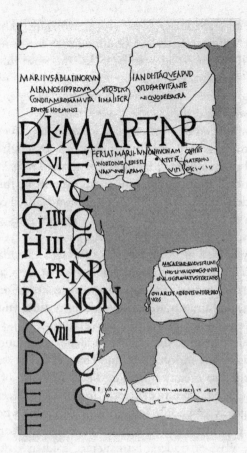

Fragments of a Roman calendar, A.D. first century, the month of March. (Letters A–H correspond to an 8-day Roman market-day cycle; K is for kalends and N for nones; to the right are fragments of holidays and historic events occurring on those dates.) Drawing by Herbert E. Duncan, Jr.

Caesar's calendar also injected a new spirit into how people thought about time. Before, it had been thought of as a cycle of recurring natural events, or as an instrument of power. But no more. Now the calendar was available to everyone as a practical, objective tool to organize shipping schedules, grow crops, worship gods, plan marriages, and send letters to friends. Combined with the rising popularity of sophisticated sundials and water clocks, the new Julian calendar introduced the concept of human beings ordering their own individual lives along a linear progression operating independent of the moon, the seasons, and the gods.

Nothing symbolized this better than a public sundial erected in 10 B.C. by Augustus to commemorate his victory over Mark Antony and Cleopatra, and to inaugurate his coming empire of peace. He used an obelisk transported from Egypt (which still stands in the Piazza del Popolo in modern Rome) as a 50-foot gnomon planted on the Campus Martius in the middle of an enormous grid of lines showing the length of hours, days, and months, and the signs of the zodiac. Beside it was probably etched in stone a copy of Caesar's calendar. "No one entering the Campus Martius could fail to see that the Caesars united heaven and earth," writes historian Arno Borst, "the Orient and the Western world, and the origin and evolution of time and history, or that they marked the beginning of a universal time."

Not that every person in the empire suddenly abandoned their age-old calendars using the moon, stars, and changes in the seasons. Only those who needed to measure time in the civic world of the empire did so. This excluded illiterate peasants, laborers, and slaves who made up the vast majority of people living inside of Rome's borders. Still, for the first time in European history the coming Pax Romana would foster a middle class of traders, bureaucrats, soldiers, lawyers, moneylenders, and craftsmen who came in contact with the notion of measuring time using numbers and calculations.

The advent of precision timing also led the Greeks and Romans to experience the first known frustrations about time as measured by a clock. Lawyers, Plato said some four hundred years before Augustus erected his gigantic sundial, were "driven by the *clepsydra* [waterclock] . . . never at leisure." Aristotle also groused about people watching the clock even during a performance in the theater. "The length of the tragedy should not be judged by the *clepsydra*," he said, "but by what is suitable for the plot." Undoubtedly many Romans felt the same way, though they must have been equally glad to have a clock when it came to setting time limits on a long-winded lawyer, or to possess a calendar to prove to one's moneylender that it had been 16 days, not 17, that one owed interest on those 10 pieces of silver.

As Caesar was making his reforms in Rome he summoned Cleopatra from Egypt. Soon after, the queen appeared with her infant son, Ptolemy XV Caesar, known as Caesarion. She also brought key figures from her court, including almost certainly the astronomer Sosigenes. Moving into Caesar's suburban mansion on the Janiculan Hill above the Tiber, Cleopatra's arrival and obvious affair with the dictator of Rome caused a scandal in Rome and "incurred the greatest censure"—not because she was his mistress but because she was a foreign monarch with a political agenda viewed as being not entirely compatible with Rome's.

Resentment grew as Caesar became increasingly aloof, forcing even powerful political leaders such as Cicero to wait for long periods of time merely to talk to him. In part this was because he was absorbed with reforms and construction projects, including plans to build a library in Rome greater than the one in Alexandria. But many interpreted his attitude as arrogance and a desire to be treated like a king—which he was in all but name, though he wisely refused the actual title, knowing that it would offend republican sensibilities in a city where becoming king was still officially punishable by death. In one famous scene involving in a small way Caesar's new calendar, Mark Antony, his loyal follower and lieutenant, placed a golden diadem on Caesar's head. When the dictator refused the crown and ordered it sent to the temple of Jupiter, Antony ordered Caesar's refusal to be king recorded in Rome's official calendar.

The outrage of Caesar's enemies rose from a simmer to a boil during the early weeks of the second Julian year, in 44 B.C., as he organized a military campaign in Parthia to begin on March 18. Just before his departure he planned to attend the Senate. Feeling ill on March 15, he arrived late by litter to the senatorial curia. On the way in he ran into an augur named Spurinna, who had supposedly warned him earlier to beware the ides of March. Caesar laughingly told the priest that the ides were here and nothing had happened. Spurinna answered that the day was not yet over.

Caesar, who had sent away his bodyguard, then moved to take his seat inside. Walking through the senators, he sat on his gilded throne and was approached by a group of lawmakers. One of them, Tillius Cimber, asked him to support a petition. When he refused, Cimber grabbed the dictator and tore the toga from around his neck. At this signal several men attacked. Caesar grabbed a dagger and was able to fend off his assailants at first. But there were too many of them; 23 daggers stabbed at him, and he fell. Bleeding to death amidst the stunned senators of Rome, this man who thought he could rule time itself drew his toga over his head and died.

4

A Flaming Cross of Gold

By the unanimous judgment of all, it has been decided that the most holy festival of Easter should be everywhere celebrated on one and the same day.

—CONSTANTINE THE GREAT, A.D. 325

Three and a half centuries after Caesar died, Flavius Valerius Aurelius Constantinus stood on a bluff above the Tiber. Kneeling down in prayer he looked up in the sky and saw a flaming cross blazing above the sun. The cross was inscribed with three Greek words: *en toutoi nika*, "in this sign conquer." That night Constantinus, whom we know as Constantine the Great, dreamed he heard a voice as he slept amidst his army bivouacked north of Rome. The voice assured him victory in a battle to be fought the next day if he marked his standard with an X cut through with a line and curled around the top: the cross symbol of Jesus Christ.

At dawn on October 27, A.D. 312, Constantine gave the order to paint the cross just before his legions attacked his chief rival at Saxa Rubra, Latin for "red rocks." Brilliantly outmaneuvering the forces of Marcus Aurelius Valerius Maxentius, who had ruled Italy as coemperor, Constantine pushed his enemy's troops into the water near the Mulvian Bridge. Earlier Maxentius had foolishly breached this ancient stone causeway, thinking it would prevent his foe from crossing. Instead Maxentius drowned along with thousands of his legionnaires, leaving the road to Rome open to the 39-year-old victor and his cross-emblazoned legions.

Whether or not one believes Constantine about his visions—even the sycophantic court historian who later recorded them expressed his doubts—his victory at the Mulvian Bridge was a crushing personal triumph and a watershed moment for Europe precisely *because* he gave the credit to the Christian god. The West would never be the same. Nor would the way people thought about time and the calendar.

Shaking up the old order was exactly what Constantine had in mind with his talk about flaming crosses and a powerful new god, though his embrace of Christianity was motivated as much by politics as by faith. It was all part of a grand strategy to forge a new imperial order—political, spiritual, military, economic.

The empire badly needed it. Racked and bloodied by almost a century of civil war, assassination, economic decline, and enemies pressing in on all sides, the Roman Empire in 312 would have been unrecognizable to Julius Caesar. Rome itself and its ancient institutions of temple and Senate had been largely eclipsed by the all-powerful imperium, a massive bureaucracy of civil servants, provincial governors, and army officers headed up by a single man—the emperor. A centralized system originally designed by Augustus, it had worked well during Rome's golden age in the first and second centuries, when powerful and relatively enlightened rulers such as Trajan, Hadrian, and the Antonines tended to occupy the throne. By the 300s, however, this old order was all but shattered, with emperors ruling at the whim of the legions, and the empire weakened by a general stagnation as it ceased to expand militarily and economically and became mired in bankrupting wars inside and out.

One alarming sign of internal decay was the sharp decline of science and the arts as the empire diverted resources to the military and people's minds became less occupied with the length of the year and poetry and more with defending their cities. In the 260s a plague exacerbated the decline, depopulating several provinces. That same decade the cities of the empire began to dismantle stone monuments and amphitheaters to erect walls against invaders. The emperor Aurelian, fearing an attack on Rome itself in the 270s, persuaded the Senate to pay for a massive new wall encircling the city.

Until Constantine's predecessor Diocletian began reasserting order in the 280s, it looked as if the empire would crack apart. In the 250s the Germanic Marcomanni crossed the Danube frontier and raided southward into northern Italy. To the east the Goths invaded Macedonia and later joined with Scythian hordes to invade Asia Minor and ransack the Black Sea coast. In 260 the emperor Valerian was captured by a revitalized Persian empire, whose armies ravaged Asia Minor until being beaten back. In 267 a fleet of five hundred Gothic warships broke out of the Black Sea and pillaged the Greek coast, sacking Athens, Argos, Corinth, Sparta, and Thebes before Emperor Claudius II defeated them in a battle that would have left Italy and Greece defenseless had he lost. All of this had left the empire unmanageable by anyone but a truly exceptional leader.

Constantine was such a man. Reigning 31 years, when most of his immediate predecessors had survived at best a few score months, this last of Rome's great emperors worked tirelessly to restructure and rejuvenate the empire, efforts that helped to fend off collapse in the West for another century and a half and in the East for over a thousand years.

Born in Naissa—Nis in present-day Serbia—this round-cheeked man with flared nostrils and a square forehead could be ruthless and never hesitated to plunge the empire into war to further his own ambitions. But he also repaired imperial highways and established an efficient messenger network, revamped the legal system, built magnificent basilicas, aqueducts, monuments, and churches, and mostly kept the peace. He also sought to transform the imperium itself, completing a shift begun by Aurelian and Diocletian toward an oriental-style monarchy where kings ruled not by the grace of the Senate and the people, or even the army, but as all-powerful despots who claimed to be chosen by the gods (or God).

Aurelian (ruled 270–275) had launched this transformation by founding a cult of a monotheistic sun god in Rome during the 270s, a precursor to the imposition of Christianity. Building a resplendent new temple to the sun in Rome, Aurelian announced that the sun god had made him emperor, not the Senate; a transformation cut short by his assassination shortly thereafter. Diocletian (ruled 284–305) furthered this eastern tilt by also embracing the cult of the sun and by dividing the empire into eastern and western halves, with the main center of power under his control in the East. He gave up the traditional purple toga of the emperor for sumptuous silk robes and jewel-encrusted belts and shoes; and, for the first time since the early days of the Latin kings, a Roman head of state donned a crown. Constantine would complete this easternification by choosing Byzantium as the site for his new capital, Constantinople. Strategically located near the empire's richest provinces, it was within striking distance of both the western and eastern frontiers.

Constantine would also adopt one of the East's chief religions, reversing 350 years of largely secular rule—symbolized by Caesar's calendar—in a move that would soon fuse the political and military might of a still-potent empire with what would become an even more potent state religion.

At first it was not entirely clear *which* religion. During these troubled times Romans embraced several popular sects, most of them from the East—everything from a pseudoreligious brand of Neo-Platonism to Christianity and the worship of the sun. Keenly aware of this diversity, the always expedient Constantine seemed willing to embrace virtually any religion that might serve his political needs, despite his story— told much later—about the Christian god and the flaming cross of gold at the Mulvian Bridge. In fact, at the time he credited his victory over Maxentius to more than one god.

To please the pagans of Rome, he erected the Arch of Constantine, which dedicates his triumph to Rome's old deities—and remains one of the best-preserved and most imposing triumphal arches in present-day Rome. Constantine also flirted with the popular cult of the sun god Mithras at the time of the battle since the Mithraists also held sacred a symbol similar to a cross. Certainly such a twin billing would have pleased the large numbers of both Mithraists and Christians in his army.

Only over the course of several years did Christianity gradually win out, perhaps because the Christians offered a more effective power base, or because Constantine found the tenets and organizational structure of the Christian Church easier to co-opt and merge into the existing imperial structure. Another, simpler reason may have involved Constantine's mother, the British-born Helena (248–328), a former barmaid and a longtime Christian who was mistress (and possibly a first wife) to Constantine's father. A formidable woman who seldom left her son's side, Helena lobbied hard for the Christian god, receiving generous sums from her son to build dozens of churches from Judea to Gaul, including the still-standing Church of the Nativity in Bethlehem and the Church of the Holy Sepulcher in Jerusalem. Constantine himself hedged on a full personal commitment to his own state religion until 337, when he was finally baptized on his deathbed.

Whatever Constantine's true personal beliefs, his fusion of church and state ended what was in essence an experiment begun by Caesar and Augustus to decouple religion from the government—and religion from time. Its impact would utterly transform Europe over the next several centuries, affecting all aspects of life, including the way people kept track of calendar days.

Inevitably Constantine's new order, like Caesar's three and a half centuries earlier, got around to putting its stamp on the calendar, in this case by creating a new, religiously inspired system of measuring time. He did this by leaving intact Caesar's basic calendar of 365¼ days and 12 months, while making three major changes within this structure: the introduction of Sunday as a holy day in a new seven-day week; the official recognition of Christian holidays such as Christmas with fixed dates; and the grafting onto the calendar of the Easter celebration, which is not a fixed date, being tied to the Jewish lunar calendar in use when Christ was crucified. The existence of these two types of holy days, fixed and floating, is where Christians get the terms "immovable feast" and "movable feast."

The emperor's first move to reorder the calendar came in an edict issued in 321, nine years after the Battle of the Mulvian Bridge, when he established Sunday as the first day in a seven-day week—a unit of time unknown in the original Roman calendar of kalends, nones, and ides.★ According to Constantine's dictate, all citizens other than farmers were ordered to abstain from work on *dies Solis*—the Sun's day. He also ordered the courts closed for litigation and the commanders of the army to restrict military exercises so that soldiers could worship the god of their choice.

Constantine's selection of Sunday was not without controversy. It blatantly rejected the long-held observance of Saturday as the Sabbath by Jews and by Roman pagans, who in the late empire had set aside Saturday—Saturn's day—as a day to rest and worship.

Saturday at one time was the choice of many Christians as well, since most early believers were Jews who felt obligated to keep their traditional holy day on this seventh day in the Jewish week. But because Jesus was crucified on the *sixth* day of the Jewish week and, according to the Bible, rose from the dead on the first day of the next week—a Sunday—some early Christian leaders decided to shift their Sabbath to Sunday, and to mark this day each week by a special service featuring the Eucharist.

But old ways died hard. As late as the turn of the second century, Christian prelates were still complaining about certain Christians who continued to favor a Saturday Sabbath, which one bishop condemned in a letter as a "superstition," describing "the show they make of the [Jewish] fast days and new moons" as being "ridiculous and undeserving of consideration."

By the time Constantine issued his edict Christians had largely set-

---

★The Romans did have an informal cycle of market days held every eight days.

tled the issue of Saturday versus Sunday, with Sunday the victor. The emperor, however, did not strike a purely Christian line with his new law. By placing the Sabbath on the day devoted to the sun in the seven-day cycle of pagan planet-gods, the emperor also curried the favor of the Mithraists and other sun worshipers. Constantine's official designation of this day in the Roman legal code as *dies Solis* cannot have pleased his new hierarchy of Christian bishops, priests, and laymen, even if some tried to justify the emperor's decision by insisting that Christ, like the sun, was the light of the world.

As for Constantine's new seven-day week, it had already been gaining in use and popularity among Romans because of its astrological significance—seven referring to the number of planets (including the sun and the moon) then thought to be in the sky, each of which "controlled" a day of the week. Indeed, the seven-day system was already ancient by Constantine's day. It seems to have originated circa 700 B.C. in Babylon, when astrologers assigned their planet-gods to the days of the week—names the Romans replaced with their own planet-gods. For instance, the day of Nabu, the Babylonian god of the scribes, became in Latin the day of Mercurius, the Roman god of communication—and today survives as *mercredi* in French, *miércoles* in Spanish, and so forth across the spectrum of Romance languages (see chart on page 56).

In English, however, the day of Nabu is known as Wednesday because of a curious twist of history: the fact that the seven-day week did not penetrate to Britain until the era of the Anglo-Saxon conquests in the fifth century. At that time the invaders wanted to take on certain Roman trappings but clung to their own pagan religion and gods. So Nabu in Babylon became Mercurius in Rome and Woden—the German (and Viking) god of poetry—in Britain. Centuries later this Mesopotamian-Roman-German-British astrological connection has spread to dozens of countries around the world, as people from Hong Kong to Harare pay homage to otherwise forgotten gods every time they mention the word *Wednesday*.

Planet	Ancient Planet-gods			Modern-day Names		
	Babylonian	Roman	Anglo-Saxon	English	French	Spanish
Sun	Shamash	Sol	Sun	Sunday	*dimanche*	*domingo*
Moon	Sin	Luna	Moon	Monday	*lundi*	*lunes*
Mars	Nergal	Mars	Tiw	Tuesday	*mardi*	*martes*
Mercury	Nabu	Mercurius	Woden	Wednesday	*mercredi*	*miércoles*
Jupiter	Marduk	Jupiter	Thor	Thursday	*jeudi*	*jueves*
Venus	Ishtar	Venus	Freya	Friday	*vendredi*	*viernes*
Saturn	Ninurta	Saturnus	Saturn	Saturday	*samedi*	*sabato*

Astrology was so influential in the ancient world that 7 became a kind of mystical number. This was evident not only in the seven-day week but also in the so-called seven ages of man. The astronomer Ptolemy, among others, believed that these ages were tied to the seven planets and their orbits in the earth-centered universe. According to his cosmology, infancy is ruled by the moon, childhood by Mercury, adolescence by Venus, youth by the sun, manhood by Mars, middle age by Jupiter, and old age by Saturn. The planets and the number seven were also associated with good and evil omens affecting winds, rain, fair sailing, good crops, bets at the chariot races, warfare and birthdays, for instance in this nursery rhyme:

> *Monday's child is fair of face,*
> *Tuesday's child is full of grace,*
> *Wednesday's child is full of woe,*
> *Thursday's child has far to go,*
> *Friday's child is loving and giving,*
> *Saturday's child works hard for its living,*
> *And the child that's born on the Sabbath day*
> *Is bonny and blithe, and good and gay.*

Recently chronobiologists have discovered that the seven-day cycle, like the sleep cycle of days and nights, may also have biological precedents. They say that certain biorhythms in the human body work on seven-day cycles, including variations in heartbeat, blood pressure, and response to infection. The potential for rejection of a transplanted organ seems to peak at seven-day intervals. Other organisms, including bacteria, share these basic biorhythms. Possibly this faint tick of biology may be one reason that Mesopotamians, Romans, and numerous other cultures, from the Incas of Peru to the Bantu of central and southern Africa, have shaped their activities around a week of 5 to 10 days.

Astrology was responsible for yet another curiosity in our weekly calendar: the order of the days. We take the order of Monday, Tuesday, Wednesday, and so forth for granted, but in fact it does not correspond to the ancient understanding of the solar system, which put Saturn farthest from the earth, followed in descending order by Jupiter, Mars, the sun, Venus, Mercury, and the moon. The discrepancy between this order and the arrangement of our week comes from another invention from Mesopotamia: the division of the day into 24 equal units of time. The reason for this scheme has been lost. As I mentioned earlier, it may have had something to do with dividing the day into two 12-hour periods to correspond with the 12 signs of the zodiac. Another reason may have been the fact that 24 works into the basic Mesopotamian numeric system based on 6. Twenty-four is divisible by 6; likewise, the 360 degrees of the Babylonian circle is divisible by 24.

The order of the day names themselves comes from ancient Mesopotamian astrologers' attaching a planet-god to preside over each hour of the day, arranged according to their correct cosmological order. For instance, Saturn controlled the first hour of Saturn's day

(Saturday), followed in its second hour by Jupiter, then by Mars, the Sun, Venus, Mercury, and the moon. In the eighth hour the cycle started again with Saturn, and the progression repeated until the twenty-fourth hour of the day, which happened to fall to Mars. Because the next hour in the cycle—the first hour of the new day—belonged to the sun god, the day after Saturday was called Sunday.

The ancients used a simple device for keeping track of the proper names of the hours and days in relation to the planet gods. They used a seven-sided figure, with each vertex marked with a planet's name in the proper order. Archaeologists found one of these wheels drawn as graffiti on a wall when they excavated Pompeii. It looks something like this:

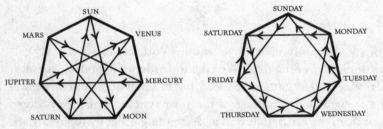

Even after Constantine's edict about Sunday, it took another generation or two for the seven-day week to catch on throughout the empire. The 24-hour system took longer, having to wait until the invention of the mechanical clock in the Middle Ages by monks anxious to observe with precision their canonical hours. Before this, people marked the passage of time during the night by using the stars and during the day either by eyeballing the sun or by listening to public announcements of the time. For instance, the Roman military had callers watching the position of the sun to announce the changing of the guard at the third hour of the morning *(tertia hora),* at the sixth of midday *(sexta hora),* and at the ninth of the afternoon *(nona hora).* In another example, Saxons in Britain divided their days according to the ocean's tides—"morningtide," "noontide," and "eveningtide."

Saxons also gave us the English word *day*, which comes not from the Latin *dies* but the word in Saxon for "to burn," during the hot days of summer. *Hour* is from Latin and Greek words meaning "season." Originally it referred to the fact that the length of the daylight period varies according to the season.

The second important calendar change introduced by Constantine was when to celebrate Easter, a matter not as easily resolved as the question of Sunday. The holiest day for Christians, Easter's worship is complicated by the fact that Christ's resurrection occurred during the Jewish Passover, which is dated according to the phases of the moon in the Jewish calendar. This means that the date for Passover—and Easter—drifts against the solar calendar, changing year to year. For early Christians this was a conundrum because they lacked the detailed astronomical know-how required to synchronize precisely the moon's phases with the solar year.

This hardly stopped Christian time reckoners from trying. Indeed, even as science and knowledge from the ancient era began to fall away in these latter days, the question of when to celebrate Easter remained one of the few areas where scientific inquiry would survive during the great darkness to come. But this was still in the future. For Constantine the issue was not so much how to determine the date for Easter, but how to get the various factions of Christianity to agree to celebrate the Resurrection on the same day, even if technically this date was not exact. Politically this was crucial to establishing one state religion, with one set of rules.

The Easter question came to a head in what is today a quiet Turkish village famous as a lakeside respite for Turks weary of chaotic Istanbul, some 80 miles away. Known as Iznik, this village 1,700 years ago was a prosperous Hellenistic city called Nicaea, Greek for "victory." This name appealed to Constantine, who styled himself "Constantinus Victorus." One historian writes, "The beautiful town lay on an eminence in the midst of a well-wooded flower-embellished country, with the clear bright waters of the Ascanian Lake at its foot." Says another, "The bright green of the chestnut woods in early summer stood out in the foreground; in the distance the snow-capped Olympus towered over its mountain ranges." It was here in 325 that Constantine convened the first major Christian council, which made the first concerted effort to solve the Easter problem and to come up with a unified date for its celebration.

The choice of Nicaea was no accident. Situated strategically in the East, near the new heart of Constantine's revamped empire, the city was easily reached by the three hundred or so bishops who attended, and their delegations. Nearly all of these came from the East, in part because Christianity had permeated few areas in the West. Sylvester I, the aging bishop of Rome—at this time all major bishops were called by the honorific *papa,* or "pope"—did not come because he was too ill, but he sent representatives.

Constantine was so anxious to convene this meeting that he paid the bishops' expenses, placing at their disposal the empire's system of public conveyances and posts along its highways. At Nicaea he paid for food and lodging. The sessions were held at a large basilica converted into a church and in the audience chamber of an imperial palace, possibly situated on the shore of today's Lake Iznik.

The council opened in the late spring, probably on May 20, without Constantine. He came a month later. The early sessions were held in the city's main church, with the doors open to the lay public. Even pagan theologians participated in some of the debates. Gathering in small groups under colonnades and in gardens, dressed in togas and robes, they argued the relationship between God and Christ and the meaning of passages in holy texts, breaking for sumptuous meals of wine, meats, fruits, and vegetables laid out by imperial servants.

For many of the bishops and priests it must have been a heady moment, if slightly surreal. Just a few years earlier many of them had been practicing their religion in secret. Some had been viciously persecuted. Paul, a bishop from Neo-Caesarea, had lost the use of his hands after being tortured with hot irons. Two Egyptian bishops each had had an eye gouged out. One of these, Paphnutius, also had been hamstrung. Constantine later singled him out at Nicaea and kissed his mutilated face. The historian Eusebius, an eyewitness at the council, writes about the lavish feast held on July 25 to celebrate Constantine's twentieth year as emperor, and the lingering fear felt by the bishops as they passed guards in the banquet halls and saw "the glint of arms" that so recently had been turned against them.

But this turnaround from fear to feasting was nothing compared to Constantine's sudden transformation of a church that for three hundred years had lacked a central authority. Scattered and at times hounded by the authorities, Christianity had operated less as a single cohesive religion than as a collection of sects and denominations following the same basic tenets but differing on points major and minor—such as when to celebrate Easter. Unity had always been a goal, though most congregations had remained more or less independent of one another, with doctrine and details of worship left to the local elders and members to decide. In cities large enough to assign a bishop, these prelates had exercised some authority, but as one historian noted in talking about Alexandria's freewheeling churches, with their many controversies and bickering between sects and church leaders, "it was not an exceptional thing to have a doctrine of one's own."

Constantine's mandate at Nicaea was to put a lid on this free-for-all by establishing a set of uniform rules governed by a centralized structure headed by himself as emperor. To accomplish this, Constantine called on the bishops to resolve differences ranging from petty disputes to fundamental controversies, the most important one at the time being the question of whether or not God the Father came before

Christ the Son, or if they both had always existed. A popular Alexandrian theologian and preacher named Arius had been espousing the
former, teachings recently condemned by his chief rival and detractor,
the bishop of Alexandria. Both Arius and the bishop had been invited
to make their case at the council.

Constantine arrived at Nicaea on about June 19, 325, and was immediately handed a thick packet of papers detailing controversies large
and small among the attendees. He carried the packet with him into
the audience hall of his palace, where he officially opened the council
wearing a robe of gold and draped with jewels like a Persian king.
Sitting on a golden throne in front of the prelates, he listened to
welcoming speeches before rising to answer the mostly Greek-
speaking bishops in Latin. Through a translator he welcomed them
but quickly got to the point about the purpose of this council, holding
up the packet of papers like a scolding father. He told them, "I your
fellow servant am deeply pained whenever the Church of God is in
dissension, a worse evil than the evil of war." Ordering the bishops
to set aside their arguments, he took the packet and dropped it into
the flames of a brazier. As it burned he told his audience that they
must use this council to establish a uniform doctrine they all would
follow—an imperative that became the guiding force behind the
Catholic ("universal") church for centuries to come and would profoundly affect all aspects of life, including attitudes toward measuring
time.

Details of the Easter debates at Nicaea are not recorded, although the
controversies leading up to the council are well known. For almost
three centuries this issue had frustrated the followers of Christ, who
were anxious to celebrate properly the signal event in their religion.

The problem arose because no one who witnessed Christ's death
and resurrection had thought to jot down a date. Even worse, the

Gospels that recount Christ's biography offer contradictory information in vague references about the timing of these events. All agree that Christ rose on the first day of the Jewish week—a Sunday. But which Sunday? Three Gospels—Matthew, Mark, and Luke—suggest the Sunday after the Passover feast in the Jewish month of Nisan. The Gospel of John, however, indicates another date in Nisan, a dichotomy exacerbated by the drift of the Jewish lunar calendar in the years following Jesus's crucifixion.

This vagueness arose because the earliest Christians cared little or nothing about dates, for the understandable reason that Jesus's disciples and first followers fervently believed in their savior's imminent return. For them time was irrelevant, a point underscored by the apostle Paul, who did not date his letters that appear in the New Testament. He explains why in an epistle written to the church in Galatia, in which he reprimands those Christians who pay attention to "days and months and times and years," accusing them of being more interested in astrology and earthly matters than in God. In another letter Paul exhorts the Christians of Colossae, in central Asia Minor, not to judge others by what they eat or drink, "or in respect of a holiday, or of the new moon or of the Sabbath, which are a shadow of things to come."

When Jesus failed to return immediately, Christians realized they needed some sort of system for dating. By the second century they started writing schedules of when to worship, and crude calendars of saint's days and other Christian holidays. They also began to argue about dates, such as whether to worship on Saturday or Sunday, and how to draw up a chronology of events in Jesus's life. This became increasingly important to a religion that is based on real events as recorded in the Bible, which says that Christ lived in actual time: he was born, raised by Mary and Joseph, was baptized, became a teacher, was tried and executed, and rose from his tomb three days later. These central events are the underpinnings of the Gospels and of Christianity itself, which makes this a religion of history and the calendar—a potent and critical reality for early adherents even as they grappled with another core tenet of their religion: the doctrine of eternal life and a God who exists *outside* of time.

This dichotomy between the Christ that exists beyond time and the historic Christ became an early source of tension in Christianity. It later became one of the great theological conundrums of the Middle Ages, when the timeless Christ of dogma and mysticism reigned supreme. Even so, the notion of empiricism and measuring time never entirely died out, in part because of the Church's need to understand enough about the temporal world to designate a proper date for Easter.

By the time of Nicaea, Christians had more or less agreed upon dates for celebrating Christ's birth and other key events. These included days set aside to mark the martyrdom of saints—dates meant to record in real time important episodes in the Christian calendar, and to provide an alternative to pagan holidays. The first known martyr's day to be commemorated seems to have occurred in the mid-second century, when the bishop of Smyrna was burned at the stake "on the second day in the beginning of the month of Xanthicus,★ the day before the seventh kalends of March, on a great Sabbath, at the eighth hour. He was arrested by Herod, when Philip of Thralles was High Priest, and Statius Quadratus Proconsul, during the unending reign of our Lord Jesus Christ." According to an eyewitness, the bishop's bones were taken away and interred in a place "where the Lord will permit us . . . to assemble and celebrate his martyrdom— his 'birthday'—both in order to commemorate the heroes who have gone before, and to train and prepare the heroes yet to come."

As for Easter, most Christians agreed by 325 that it should be preceded by a fast, and that the sacred day itself should have some relationship with the full moon that falls during the Jewish month of Nisan. Beyond this, individual churches and sects split on the issue of holding Easter always on a Sunday or according to the approximate date in Nisan that Christ rose from the dead, which changed according to the drift of the Jewish lunar calendar. By the third century a rising

---

★A month in a local Greek calendar.

anti-Semitism among non-Jewish adherents added to the confusion, as Christians became biased against using dates that depended on when Jewish priests determined the start of Nisan to be. So a third choice emerged: linking Christ's resurrection to the solar year and to Caesar's calendar by using the spring equinox as a fixed astronomic date to determine Easter. With this anchor date decided, a formula could be devised to correlate the equinox with the phases of the moon and the weekly cycle of Sundays.

None of the surviving canons issued by the council mentions the Easter problem directly, though the rules that emerged from Nicaea are well known among Christians: that Easter will fall on the first Sunday after the first full moon after the equinox, but shall never fall at the beginning of the Jewish Passover. The sentiment of the assembled bishops was recorded by Constantine himself in a letter addressed to bishops and other church leaders who did not attend the council. "By the unanimous judgment of all," wrote the emperor, "it has been decided that the most holy festival of Easter should be everywhere celebrated on one and the same day." In the same letter Constantine notes that the council opposed the practice of following the Jewish calendar to determine Easter. "We ought not," he says in a letter charged with anti-Semitism, "to have anything in common with the Jews, for the Savior has shown another way."

But the council's solution was hardly perfect. First off, it codified a holiday that changes dates every year, a confusing notion for the average Christian or recently converted pagan used to annual holidays falling on the same date every year. A second problem was that Nicaea's Easter solution required what was then impossible: an accurate determination in advance of a date that assumed a precise knowledge of the movements of the sun, earth, and moon. Ancient scientists could calculate only an approximate date, a reality that would haunt time reckoners for centuries as they tried, and failed, to determine true dates for Easter. In the absence of good science most churches fixed on an arbitrary date for the vernal equinox on March 21.

Another blemish in the Nicaea solution was the failure of the council's bishops and time reckoners to correct the central flaw in Caesar's calendar: the annual error of 11 minutes. This meant that an Easter tethered to a fixed spring equinox would drift backward with the rest of the calendar, falling behind the true orbit of the earth by one full day every 128 or so years. By 325 the Julian calendar was already three days behind where it stood when Caesar introduced his reforms in 45 B.C., when the vernal equinox fell on March 25. By Bacon's day the true equinox had dropped back to March 14, though the church continued to follow the practice after Nicaea of rigidly determining Easter according to a March 21 equinox, arbitrarily set at the time of the council.

On the other big issue at Nicaea—the nature of Christ—the council debated heatedly throughout that long-ago summer, finally issuing on July 25 the Nicene Creed, which declared Arianism a heresy and affirmed that Christ and God came from one substance and had both always existed. But far more important than the nature of Christ or the date for Easter was Nicaea's codification of Constantine's fusion of church and state, an expedient political move by this shrewd emperor that was to link inexorably the Church to secular power, wealth, and absolutism for many centuries to come—first as an adjunct to imperial Rome and later as an independent entity that derived its all-embracing influence from its own imperial-style hierarchy and assumption of power over Christian domains.

Constantine closed the council by admonishing the still-fractious bishops to keep their unity at all costs and to use their newfound power with care. "Be like wise physicians," he said, "who treat different cases with discrimination, and are all things to all." Undoubtedly no one assembled on that hot Mediterranean day, feasting on Constantine's meats and fruits and sipping his wine, had any idea how prophetic the emperor's final words would be—that this recently outlawed religion would truly become "all things to all" in every realm, including time, replacing Rome itself as the most powerful single entity ruling the lives and souls of millions of people and countless generations yet to come.

5

Time Stands Still

Try as they may to savor the taste of eternity, their thoughts still twist
and turn upon the ebb and flow of things in past and future time. But
if only their minds could be seized and held steady, they would be still
for a while and, for that short moment, they would glimpse the splendor
of eternity, which is forever still.

—AUGUSTINE OF HIPPO, C. A.D. 400

Less than a century after Constantine celebrated the success of his council at Nicaea, a Roman foot soldier stood sentry on a snow-swept river bank at Mainz in what is now Germany. Shivering in his armor and military wraps, this anonymous infantryman watched the ice-choked Rhine and the opposite bank, where hundreds of cooking fires burned, tended by a vast and growing horde of German *barbari*. This lone soldier might have been a Roman, or more probably a Romanized German recruited by the faltering empire to help defend its northern border. Whatever his nationality, as he stamped his feet to stay warm on this frigid December day in 406 he almost certainly was not thinking that Rome itself hung in the balance. Even when he looked up and saw to his horror that the masses across the river were moving toward him over the ice, he could hardly have imagined that this was the beginning of the end of the ancient world in the West, and for Europeans an end to time itself as they had known it.

The sentry sounded the alarm and his legion scrambled to meet the barbarians, a coalition of tattooed, scraggly, fur-clad Germans from the tribes known as the Alans, the Sueves, and the Vandals. But the

Roman garrison was fatally depleted. Most of the men had been pulled off the Rhine frontier to join a desperate counterattack against yet another army of *barbari*, the Ostrogoths, then invading the Balkans. Removing the Rhine legions was a calculated move by the Roman military, who were betting that the Germans there would not attack during the winter. But no one had counted on the fact that the Rhine might freeze solid, a rare occurrence. Nor could the emperor and his generals have known that the Germans were themselves fleeing the savage invasion of their country by the Huns.

Lacking the resources to stop them, Rome watched helplessly as the Mainz hordes and other waves of invaders poured across borders that had held firm for four hundred years to ravage defenseless cities. Britain was lost in 410 when its Roman garrison departed to defend Gaul, never to return. Soon after, Gaul itself began to break apart; Spain too slipped slowly away in the West, along with parts of the Balkans in the East. A marauding band of Visigoths reached the gates of Rome itself in 410, crashing through its walls to sack a city that for centuries had been one of the greatest powers in the earth's history.

Inevitably the gathering chaos affected people's perceptions of time and the calendar as the predictable patterns of Roman life began to crumble. Caesar's calender would remain the official calendar in the West long after the empire fell, though more and more people found an organized list of days, months, and years irrelevant. They had more immediate concerns, such as finding enough to eat and avoiding the ravages of the *barbari*.

But chaos was not the only outcome of the empire's collapse. Nor did every Roman institution falter. One, in fact, grew stronger amidst the disorder and decay: the Catholic Church. Originally designed by Constantine as a vehicle to enhance the political might of Rome, the church ended up superseding it, retaining its power and influence in the ecclesiastic realm, particularly as the *barbari* dropped their pagan gods and embraced a Church that demanded—and got—an allegiance much stronger than the imperium itself had ever known. This was because the church claimed jurisdiction not over lands and armies but over souls, an authority that would extend during the coming centuries into virtually all aspects of a Christian's life.

This amounted to a new societal order in Europe, including a new concept of time—something Christian theologians call *sacred time*. Neither cyclical nor linear, it is rather a kind of antitime that Christians equate with God, who is perfect, eternal, and timeless.

The idea of sacred time was hardly new. In one form or another it had existed since religions developed concepts of eternity and the afterlife, core beliefs for ancient Egyptians, Jews, and many other cultures. Sacred time had been a part of Christianity from its earliest days, though much as they do today, Roman-Christians had tended to keep God's time in their religious lives, while continuing to operate in their daily lives on real time—on the passage of hours, days, months, and years. But as Rome's political power ebbed and the Church rose from its ashes, the sacred soon overwhelmed the profane.

The man who best articulated this new order was Augustine of Hippo (354–430), a bishop and theologian who wrote two of the most influential Christian books outside of the Bible: *The Confessions of St. Augustine* and *The City of God*. In both works Augustine takes some pains to explain "sacred time" and why he believed it was more "real" than secular time, which is fleeting.

Augustine's long life straddled the years when Rome slid from a still-formidable empire under Constantine's immediate successors into the widening abyss of final decline. He was 52 years old in 406, when the Mainz hordes broke through the frontier, and lived to see the dismemberment of Gaul, Spain, and North Africa. Indeed, the backdrop of the empire's slow collapse obviously influenced Augustine's philosophical outlook, which favored a secure, perfect "city of God" over the faltering "city of man."

Born just 17 years after Constantine's death, Augustine grew up in the small provincial city of Tagaste, 40 miles from the coast of what is today Algeria. In a meteoric early career as a philosopher and teacher, he moved from his little town to Carthage, then to Rome, and finally in his early thirties to the imperial court at Milan, at that time the de facto capital of the western empire. This was during the reign of Theodosius I (d. 395), the last powerful emperor to reign over the entire empire. In his palace the young Augustine became the court teacher of rhetoric, a coveted position that might have led to high political office, power, and wealth.

But Augustine was a troubled young man. Living a life he describes in his *Confessions* as one of near debauchery and moral vacuousness, he tried and rejected several of the religions popular at the time. Then in 386, at the age of 31, he was alone in a Milan garden when he says he heard the voice of a child when no child was there. The voice commanded him to open a nearby Bible, which told him to give himself over to Christ. He did, resigning his post in the imperial household and eventually returning to North Africa to become a bishop of the small port city of Hippo—today's Annaba in modern Algeria, on the sea near the border with Tunisia.

Known as the last great intellectual of the classical era, Augustine set out to create a philosophical structure that linked his new religion to one of the giants of the ancient world, Plato, equating this long-dead Athenian's ideas about a prime mover/creator with the Christian God, and Plato's notion of a perfect universe, existing beyond our flawed world, with the Christian concept of heaven. Augustine borrowed from Plato's conception of time as being by definition in motion. This makes it an imperfect attribute of an imperfect world, since the realm of the prime mover is a place of perfection that by its nature is timeless and immutable. It has no beginning or ending, nor any movement forward or backward, and therefore has no time to measure. Recast in Christian terms, this ideal is what Augustine meant by sacred time.

"The world was made not in time," Augustine says in *The City of God,* "but together with time." This means that God the creator may have set in motion the idea of time as perceived by humans, but he himself exists outside of it, a concept that Augustine argues is ultimately a matter of faith. "Follow the One," he says, "forgetting what is behind, not wasted and scattered on things which are to come and things which will pass away . . . and contemplate Thy delight which is neither coming nor passing."

A discussion of Augustine's ontology may seem a bit abstract for a book about little squares marching along on a calendar, except that it represented a powerful current then forming in Europe and in the Church, which for centuries would cast a suspect eye on anyone who tried to delve too deeply into matters of time. Augustine understood the need for a simple calendar that kept track of holidays, legal days, and birthdays. Nor did he oppose a philosophical discussion about the nature of time. What he opposed was an overemphasis on trying to quantify the past, particularly on issues such as the creation—something he considered a waste of time for those seeking the perfection of God. He was even more critical of those who tried to predict the future, which in his mind was sole province of God. These included astronomers and mathematicians who used planets and other cues from nature to predict the future beyond the next harvest or the seasonal coming of winter and spring. "In the Gospel we do not read that the Lord said: I am sending you the Holy Spirit so that he can teach you about the course of the sun and the moon," Augustine wrote in a 404 letter. "He wanted to make Christians, not mathematicians."

Augustine, however, was hardly the last word on how to treat the past and future, and time itself. Indeed, his mysticism and reliance on faith would continue to bump up against those who wanted to categorize and measure the past—especially the Christian past—and those who wanted to plan or to predict the future in a systematic and scientific manner. It was the tension between these two ideals, the sacred and the profane, that would dominate the next millennium in Europe, though one side clearly was the victor—even as Rome's political and cultural collapse combined with Augustine's philosophy of antitime to all but extinguish any scientific interest in the calendar, or in making it more accurate.

And yet, as we shall see, the light of scientific curiosity was never quenched entirely. Even in the darkest days after the fall of Rome a progression of isolated monks and thinkers remained inquisitive, inasmuch as they were able, about nature and science—including ways to better measure what Augustine said was unmeasurable: time.

Augustine himself conceded that time reckoning could be tolerated in one area where the sacred and the profane could not be disentangled: calculating and predicting the date for Easter. This could be determined only by someone knowledgeable in astronomy and mathematics—and so the calculation of the date of Easter became the slender thread that science would hang by over the coming centuries. This was ironic, given that Christians who condemned science as a blasphemous intrusion into God's domain were forced to rely on science to date the most mystical event in their pantheon of miracles and otherworldly epiphanies: the resurrection of Christ.

The history of science in the Middle Ages would have been very different if the bishops at Nicaea had decided simply to name a fixed date for Easter in the solar calendar. But they did not. Instead, in the wake of Nicaea, Christians developed what became a complex equation to determine the proper day, forcing time reckoners to return to something Caesar had dispensed with centuries earlier: a dependence on the moon. Almost by accident they found themselves confronting the ancient conundrum of trying to correlate the phases of the moon with the orbit of the earth—the same problem that had plagued calendar makers from China and Babylon to preimperial Rome as they tried to fuse a 354-day lunar year with a roughly 365¼-day solar year.

Even today this lunar-solar linkup is a challenging astronomical problem, one that must compensate for a complicated range of gravitational tugs and pulls from the sun, moon, and other celestial bodies; the slow degradation of the orbits of the earth and moon over time; the slightly elliptical orbits of the earth and moon; and the spin of the earth on its axis—all factors that Christian time reckoners in the era of Nicaea had no inkling about when they devised their basic formula for Easter. Below is a 14-step algorithm devised by modern-day Catholic astronomers, who factor in some of the variables to come up with an almost precise Easter date—*almost,* because there are always minute fluctuations in the movements of the earth, moon, planets, and stars that make an absolutely exact measurement impossible to predict.

a = $year \% 19$
b = $year / 100$
c = $year \% 100$
d = $b/4$
e = $b \% 4$
f = $(b + 8) / 25$
g = $(b - f + 1) / 3$
h = $(19 \star a + b - d - g + 15) \% 30$
i = $c/4$
k = $c \% 4$
1 = $(32 + 2 \star e + 2 \star i - h - k) \% 7$
m = $(a + 11 \star h + 22 \star 1)/451$
Easter month = $(h + 1 - 7 \star m + 114)/31$ [3 = March, 4 = April]
p = $(h + 1 - 7 \star m + 114) \% 31$
Easter date = $p + 1$ (date in Easter month)

$/$ = division neglecting the remainder
$\%$ = division keeping *only* the remainder
$\star$ = multiply

As far as anyone knows, the bishops at Nicaea did not officially assign anyone or anyplace to make the official Easter determination, though the task naturally fell to Alexandrian astronomers. Even before the great council the bishops of Alexandria had dispatched letters to other churches announcing the date when they would celebrate the Easter feast. Few details are available about these early calculations, though the Alexandrians before and after Nicaea apparently used the old 19-year cycle of lunar months—the Metonic cycle—to link the moon to the solar year.

The Alexandrians also seem to have been the ones who fixed the date for the spring equinox on March 21, a change from Caesar's day, when the equinox was set on March 25. This shift may have been an attempt to compensate for the drift in Caesar's calendar against the true solar year, though the true drift between Caesar's reform in 45 B.C. and the Council of Nicaea in A.D. 325 was closer to three days than to four.

At least two astronomers are known to have created time charts predicting future dates for Easter. Both were also bishops of Alexandria—Theophilus (bishop 385–412), whose tables covered a 100-year span between 380 and 480, and his nephew Cyrillus, who succeeded his uncle and devised a 95-year table covering the Easters between 437 and 531. Both charts were reasonably accurate, though they suffered from a small flaw in the Metonic cycle—the fact that 235 synodic lunar months do not fit exactly into 19 Julian years, falling one day long. Over the course of 95 years (five 19-year cycles) this excess of a single day amounts to a five-day mistake in matching up the phases of the moon with the Julian calendar—a problem early time reckoners attempted to deal with by intercalating a day into each 19-year cycle.

A more serious problem for Easter reckoners after Nicaea was political rather than scientific. Not every city went along with the Alexandrians' methods for dating Easter, despite the council's dictate that the Easter question should be addressed uniformly for all Christians.

The most pronounced difference was between the churches of the East, which followed Alexandria's lead, and the churches of the West, which looked to Rome—a split that went far beyond issues of Easter and the calendar as the Roman world slowly divided itself along a fault line of east and west, Greek and Latin, Hellenistic and Roman. The Easter differences between Rome and Alexandria were small but important, particularly because they foreshadowed the eventual split between the Greek and Latin churches, which to this day celebrate Easter on different dates.

The first east-west Easter squabble concerned dating the equinox. The Egyptians continued to use March 21. Rome, however, used Caesar's original date: March 25. The other problem involved methods for matching up the solar year and the phases of the moon. Romans used a system developed in the mid-third century based on an 84-year cycle of lunar months divided into years, which was accurate within a day and a half. This differed from the Alexandrians' 19-year cycle, which was both more precise and easier to keep properly adjusted.

In most years the result of these subtle differences meant nothing, since both methods came up with the same day for Easter. A few years, however, were wildly off. For example, in 387 Augustine noted angrily in a letter that the Alexandrians were celebrating Easter on April 25 and the Romans on April 18. Worse still, he fumed that the Arian churches of Gaul—still thriving despite Nicaea's condemnation of their founder's doctrine—had come up with a *third* date. Using yet another formula, they celebrated Easter that year on March 21.

Dissension over details in the Easter calculation was one reason why Augustine at times became impatient with mathematicians and others who seemed obsessed with numbers and with measuring time. The bishop of Hippo had little patience with such worldly minutiae as he went about completing the process set in motion by Constantine and the bishops at Nicaea to subjugate time to God, and by extension to the Church. Christians had long been thinking this way, but not until Augustine did anyone lay it all out and elevate the issue of God's time from the simple language and logic of the apostles to the high scholarly realm of philosophy in the ancient tradition, an intellectual legitimization that the church had lacked before.

As the hordes of *barbari* swarmed across Gaul and Iberia in the years after the Mainz invasion of 406, tribes fanned out to plunder in every direction. One band of Vandals marched all the way from their homeland in modern Hungary to cross the Strait of Gibraltar in 429, when they began terrorizing the provinces of Mauretania and Numidia, finally reaching Augustine's city of Hippo in 430.

The aging bishop, then 75 years old, joined in the collective effort to organize the city's defenses and to care for the thousands of refugees from other Roman towns pressed inside the city walls. By midsummer Hippo was completely surrounded by the *barbari,* who set up a fourteen-month siege. Inside the walls the people grew hungry as the Vandals hemmed them in from land and sea. They then became sick with an illness that spread quickly through the crowds living in makeshift, unsanitary conditions. Stricken with fever, Augustine himself was sent to bed sometime in August. He died a short time later, several months before the invaders conquered the city, which Rome was forced to cede to the Vandals along with Carthage eight years later in a desperate gambit to appease these *barbari* before they captured other key African provinces supplying grain to Italy.

The Vandals, reveling in their plunder as they moved into the shattered cities of Roman Africa, formed a poignant backdrop to Augustine's death. For as he died the ancient world of Caesar, Augustus, and Constantine was also dying, as was time as it had been understood in ancient times.

But time did not entirely stop—not yet, anyway—despite the empire's final demise in 476 with the assassination of its last emperor, Romulus Augustulus, 46 years after Augustine's death. For even as Rome's invaders fought over their spoils, a brief and improbable window opened up in Italy late in the fifth century: a moment of peace and political stability that allowed three remarkable scholars living in

Rome to flourish in what was truly the last gasp of the ancient world. Each in his own way affected time, the calendar, and how people would perceive them in the dark ages fast approaching. Two of them were sons of ancient patrician families in Rome, young intellectuals who experienced meteoric careers as scholars and political appointees. The other was a Scythian monk and theologian about whom little is known.

By the time these three young men were living in Rome, at roughly the turn of the sixth century, the city had again changed hands. Just a few years earlier the German general Odoacer, who deposed Romulus Augustulus, was himself ousted and killed by the Ostrogoths in 493. Meanwhile Gaul had cracked into shifting territories fought over by German warlords leading bands of Burgundians, Franks, Alemanni, Alemanes, Goths, and Suessiones. In Britain bands of Picts, Angles, and Saxons fought each other as a few surviving enclaves of Romano-Britons grimly hung on, pushed west into modern-day Wales. To the south the Visigoths seized all but the far west of Iberia; in North Africa the Berbers and Vandals controlled the entire coast and the waters of the western Mediterranean with a fleet built by the Vandal king Gaiseric in Carthage. In the East the old empire persevered, but barely, getting some breathing space early in the sixth century when invading Persians, who had nearly crushed them, had to break off their conquest to beat back the Huns ravaging their own northern and eastern frontiers.

Rome itself was a shattered city, sacked repeatedly over the previous century. By now the great buildings, homes, and monuments were mostly stripped of precious metals. Basilicas, massive baths, and the labyrinth of palaces on Palatine Hill were still in use but decaying as a depleted civil service struggled to maintain what they could. Statues lay smashed on empty streets, and entire neighborhoods fell into ruin as most people abandoned the city. Markets lacked the grain and produce supplied for centuries from colonies now lost. Romans facing one blow after another had begun a centuries-long process of peeling off the marble facades and dismantling the stone from one building after another to use in new construction or to build defenses. Only those basilicas and temples taken over by the Church remained more or less untouched, though most prelates bore little regard for the art and architecture of pagans. One can still see divots on columns outside temples converted to churches, gouged by Christians who wrapped chains around them and tried to pull down these old "pagan" structures, but could not because they were built too well.

Disastrous decline seemed inevitable in Italy, as it did elsewhere in the former empire, until the arrival of an unexpected savior in the guise of King Theodoric, whose powerful Ostrogothic army had swept in from the east to take over Italy and parts of what is now France, Austria, and the Balkans. An unusually enlightened leader and clever military strategist, Theodoric ruled Italy for 33 years, providing stability for the first time in a century with a combination of a powerful army and restoration of the old imperial civil structure. A great admirer of Roman culture, Theodoric, ruling from Ravenna, capital of the last few emperors in the West, set about repairing and rejuvenating what he could of Italy's battered cities. In Rome he rebuilt palaces, shored up roads, and reopened aqueducts destroyed by the *barbari*. It was during this brief and all too furtive flash of Roman renaissance that our three young men were able to pursue political careers and intellectual pursuits, including work on time and the calendar, almost as if the old empire had not died.

The most famous of the three is Anicius Manlius Severinus Boethius, born in Rome in 480 to an ancient noble family. His ancestors included numerous consuls and senators, two emperors, and a pope. Orphaned young, he was raised by another ancient noble family headed by Quintus Symmachus, consul in 485 and later prefect of Rome under Theodoric. By 510 the 30-year-old Boethius was accomplished enough as an intellectual and politician for Theodoric to tap him for consul and for several delicate diplomatic missions—including the delivery of a water clock and sundial, long symbols of learning and of Roman culture, to the king of the Burgundians. Soon after, Theodoric appointed him to the high office of *magister officiorum,* a kind of royal chief of staff in charge of the civil service and palace officials. In 522 Boethius was honored again by the king's appointment of his two sons to the consulship, approved by both Theodoric and the emperor in Constantinople, who retained a titular authority over such offices.

But Boethius's true love was learning. This was his *summum vitae solamen,* his chief solace in life. Somehow finding the time, he plunged into scholarly pursuits, translating into Latin, Gibbon tells us, "the geometry of Euclid . . . the mechanics of Archimedes, the astronomy of Ptolemy, the theology of Plato and the logic of Aristotle." These translations are the only reason many of these works were preserved into the Middle Ages. Boethius also penned tracts on theology and a treatise on mathematics—a compendium of the knowledge of numbers that became a textbook for scholars in the Middle Ages, used by time reckoners among others, who were indebted to Boethius's careful recitation of mathematical concepts such as whole numbers, geometric equations, and proportions.

But by far Boethius's most important—and haunting—work was his thin *Consolation of Philosophy,* written while he was imprisoned in a fortress tower in Pavia by Theodoric and tortured daily during the winter of 524–525. Why the king arrested his brilliant *magister officiorum* is not clear, though historians surmise the king suspected Boethius of conspiring with the emperor in Constantinople, possibly over religious matters. Because Theodoric and the Goths were Arian, tensions inevitably ran high at times, particularly after the prelates of Constantinople and Rome settled a series of long-standing disputes shortly before Boethius was arrested. This rapprochement between the defunct western and still viable eastern wings of the old empire undoubtedly made Theodoric uneasy as the Byzantines stirred to life militarily under their new emperor, Justinian (483–565)—who in fact would invade Italy and crush the Ostrogoths a few years later.

Whatever the reason, Boethius's cruel imprisonment provides a tragic but poetic coda on the ancient world, including a farewell to the ancient view of time as something to be studied and contemplated instead of shunned or left to a few expert monks assigned the task of determining Easter. One can feel the anguish of this man, whose own imprisonment becomes a metaphor for the end of learning even as time slows down and the world, in his view, darkens:

> *So sinks the mind in deep despair*
> *And sight grows dim; when the storms of life*
> *Blow surging up the weight of care,*
> *It banishes its inward light*
> *And turns in trust to the dark without.*
> *This was the man who once was free*
> *To climb the sky with zeal devout*
> *To contemplate the crimson sun,*
> *The frozen fairness of the moon—*
> *Astronomer once used in joy*
> *To comprehend and to commune*
> *With planets on their wandering ways.*
> *This man, this man sought out the source*
> *Of storms that roar and rouse the seas;*
> *The spirit that rotates the world,*
> *The cause that translocates the sun*
> *From shining East to watery West;*
> *He sought the reason why spring hours*
> *Are mild with flowers manifest,*
> *And who enriched with swelling grapes*
> *Ripe autumn at the full of year.*
> *Now see that mind that searched and made*
> *All Nature's hidden secrets clear*
> *Lie prostrate prisoner of the night.*
> *His neck bends low in shackles thrust,*

And he is forced beneath the weight
To contemplate—the lowly dust.

In the *Consolation* Boethius finds comfort in his intellect, in striving for truth through philosophy, and through God. Indeed, his spirit, which was so clearly at odds with the anti-intellectualism then spreading over Europe, would also console those solitary monks and thinkers left tending the dim flicker of light that constituted learning through the long, dusky centuries to come.

It fell to the second of the three men in this odd Gothic-Roman world to carry on Boethius's ideals as darkness truly fell. Flavius Magnus Aurelius Cassiodorus was born about 490 to another of Rome's influential patrician families. The son of a praetorian prefect of Rome under Theodoric, Cassiodorus became his father's aide in his late teens or early twenties, while plunging into the same sort of intellectual pursuits as his friend Boethius. And like his friend, Cassiodorus was noticed at a young age by the German king, who moved him rapidly through the ranks of the imperial civil service. In 523 Theodoric appointed him to be Boethius's replacement in the top job of *magister officiorum* in Ravenna even as his friend was being tortured in prison and penning his soulful *Consolation*. It seems that Cassiodorus either did nothing to help, or could do nothing. Official correspondence preserved from this period, written mostly by Cassiodorus, does not mention Boethius's plight.

Apparently Cassiodorus was less threatening to Theodoric than Boethius was. He not only survived the immediate peril of the king's wrath, but also lived well beyond Theodoric's reign. He died decades later, long after the Goths themselves were driven out of Italy by Justinian, the Byzantine emperor who tried—and failed—to revive the empire in the West. While the Goths reigned in Italy, Cassiodorus served as a high official for Theodoric and his successors, including a daughter named Amalasuentha, who ruled eight years as regent for her infant son. For 15 years he also stood behind many of the efforts to revive and repair the crumbling cities of the Roman heartland, penning edicts in the king's name—including orders to restore and preserve monuments, an increasingly hopeless task after Theodoric's death. "Do not let these images perish," he pleaded in one edict, referring to the deterioration of certain bronze elephants on the Via Sacra in Rome, "since it is Rome's glory to collect in herself the artisan's skills whatever bountiful nature has given birth to in all the world." He also published numerous works, including a history of the Goths and a twelve-volume set of his official correspondence as *magister officiorum*—highly literate epistles that discuss a number of scientific topics, including expositions on the months of the year.

Soon after the capture of Ravenna by Justinian in 540—which briefly reconnected parts of the old western empire with the eastern part—Cassiodorus traveled to Constantinople, plunging into the intellectual life of the Byzantine capital. He stayed for a decade and a half at what was then a crossroad of old-world culture and learning and Christianity, returning home to Italy in 554. What he found there was chilling—a homeland shattered after the final convulsive wars between the Goths and Byzantines. Huge swaths of the countryside lay wasted. The city of Rome itself was virtually in ruins. In the end Justinian had won against the Goths, but the price had been the near destruction of Italy. Moreover, the Byzantines were stretched so thin that they soon would lose much of their bitterly won territory to the Lombards, yet another tribe of German *barbari* pressing against the northern frontier of Italy.

This was a critical moment for Cassiodorus and many other Roman scholars and nobles now faced with the undeniable end of the old world. They could think of only one thing to do: withdraw from the broken walls and ravaged streets of Rome and other cities to their estates in the country, which over the years of turmoil Rome's powerful families had fortified with stout walls and defenses in what became early prototypes for medieval castles. But when Cassiodorus joined the exodus to the countryside he took with him his thirst for knowledge, turning his family estate near the toe of Italy's boot into a combination school and religious retreat—a scholar's monastery, a place of learning that mixed rhetoric, mathematics, time reckoning, and other elements of a classical curriculum with religious study. In this way Cassiodorus turned his back on the outside world he had served for so long, withdrawing intellectually and spiritually as well as physically to become a spiritualist and a *conversus*—one who "converts" from a life of evil to one of living according to Christian principles.

This approach was considerably different from the majority of monasteries and communities of monks then forming in Italy and across Europe, most of whom specifically avoided any knowledge not directly applicable to their faith, or they took a stance that all useful knowledge had already been written down, so there was no use searching for more. Cassiodorus embraced both ancient and Christian thought, insisting that the monastery should be a place to worship and to preserve a spirit of learning—which included a somewhat desperate attempt by Cassiodorus to save ancient manuscripts as city libraries and schools were ransacked and abandoned.

Already in his sixties when he became a full-time monk, Cassiodorus devoted the remaining years of his long life to building up his monastery. He assembled a collection of ancient texts that some say numbered in the low thousands, and he wrote about a wide range of subjects—including a defense of old-world science that echoes Boethius's devotion to philosophy. In doing so he helped preserve the rudiments of time reckoning during the dark centuries to come, leading up to Roger Bacon's strident restatement eight centuries later of Cassiodorus's belief in the truth of science as an expression of God's creation. Around 550 Cassiodorus wrote a defense of mathematics and how it is critical to astronomy and time reckoning:

> It is given to us to live for the most part under the guidance of this discipline [mathematics]. If we learn the hours by it, if we calculate the courses of the

moon, if we take note of the time lapsed in the recurring year, we will be taught by numbers and preserved from confusion. Remove the *computus* from the world, and everything is given over to blind ignorance. It is impossible to distinguish from other living creatures anyone who does not understand how to quantify.

Cassiodorus was hardly a secularist, however. In extensive writings about arithmetic, astronomy, and the science of time reckoning—which he called *computus*—he makes a critical distinction between time *measurement* and time *reckoning*. The first, he said, is merely a matter of making observations of celestial bodies and jotting down numbers, and using mechanical devices such as clocks that require technical skill but not intellectual achievement to manufacture. Time reckoning, on the other hand, is purely intellectual, says Cassiodorus. It recognizes God's miracles of numbers and their usefulness in making *calculations* of time, which are critical to a believer for planning when and how he will worship God, with the ultimate calculation being the true date of Easter.

This did not mean that Cassiodorus disapproved of astronomy or of clocks. Years earlier a more secular Cassiodorus had written his friend Boethius that the *horologium*—a combination sundial and water clock—was the highest achievement of civilization, held in awe by barbarians. He still believed this late in life when he told his monks:

> We do not want to leave you in ignorance of hour-measurements; they were, as you know, invented for the great benefit of humanity. For this reason I had two clocks made for you, a sundial fed by sunlight, and a waterclock giving the number of hours constantly, by day and night.

But Cassiodorus did not teach his pupils how to construct these mechanisms, believing that monks should contemplate theory and calculations and not spend their time like village mechanics tinkering with devices. In this spirit the elderly Cassiodorus and his followers used their science of *computus* to create daily, weekly, and monthly calendars of sacred days and monastic duties and feasts. They also wrote the first textbook explaining how to compute Easter, beginning with the year 562—a set of instructions widely used in the Middle Ages, though not exactly as Cassiodorus intended. Indeed, for a man setting out to preserve knowledge and to inspire intellectual thought, his textbook allowed generations of monks simply to follow a cookie-cutter recipe for determining dates, rather than learn the processes behind the calculations. Likewise, Cassiodorus's water clocks quickly fell into disrepair after the master died, since no one knew how to fix them.

But it was a sign of these tumultuous times, when monks seemed to be setting up monasteries on every rocky hill in Italy, that what one monastic teacher was condemning, another was condoning. So even as Cassiodorus's clocks stopped in southern Italy another leading monastic figure not connected to our three young men in Rome, Abbot Benedict of Nursia in Umbria, was energetically teaching his monks to make clocks and to use them to tell time down to the hour— something no one had done before in such a systematic or official way.

Benedict was a typical monastic in his belief that devotees should concentrate on the hereafter, and that man's time on earth was ephemeral. But he also shared the ascetics' obsession with following rules to reinforce his faith, which led him to embrace clocks as instruments that could serve man in his service to God. In about 540, the year Ravenna fell to Justinian and Cassiodorus moved to Constantinople, Benedict wrote a guide to what he considered proper worship, known as The Benedictine Rule. This included a table of hours setting out a strict list of duties, prayers, mealtimes, and ceremonies linked to a careful measuring of each hour of the day.

Before the Rules a monastery's abbot typically arranged tasks and schedules for his tightly knit community. But Benedict, working in the spirit of creating uniform rules for the universal (Catholic) church, refused to leave this to the whim of individual abbots. Wanting to be sure that a monk in Naples was saying the same Psalm at the same hour as one in Provence, he ordered that time be kept accurately and objectively by using the best clocks then available: the sundial and water clock, and later a "candle clock" made to burn in measured hourly increments.

Benedict's Rules started with the Christian calendar as it then existed, with its saint's days, holy days associated with Christ's life, celebrations, and feasts. He then assigned tasks and duties to virtually every day of the year, using as his inspiration the Roman army's system of loosely dividing the day into hours, with daily watches rotating on the third, sixth, and ninth hours (morning, noon, and afternoon). Benedict ordered these three key points announced each day in the monastery. He also delineated canonical hours that did not have to be announced: dawn *(matutina)*, sunrise *(prima hora)*, sunset *(vespera)*, and the coming of complete darkness at night *(completorium)*. He listed certain Psalms to be read each day and at the beginning of the seven named hours so that everyone would know the correct hour and when it began. He fixed precise hours for waking, eating, working, and resting, and staggered them according to the seasons. For instance:

> During the winter, that is from 1 November *(a Kalendis Novembribus)* till Easter, the time of rising will be the eighth hour of the night, according to the usual reckoning. From Easter till 1 October *(Kalendas Octubres)* the brethren should set out in the morning and work at whatever is necessary from the first hour till about the fourth. From the fourth hour until the Sext they should be engaged in reading. After the sixth hour, and when they have had their meal, they may rest on their beds in complete silence. . . . The None prayers should be said rather early, at about the middle of the eighth hour, and then they should work again at their tasks until Vespers.

Benedict's system meant that Christian monks for centuries would live under Rome's civil calendar and the Roman army's day, imposed far more strictly than by the old empire's magistrates and generals. But the idea here was not temporal power or political order but a test of willpower and belief, and a means by which monks could fill their days with manual work that would keep their minds sharply focused on spiritual matters. "Idleness is an enemy of the soul," wrote Benedict.

The abbot of Nursia's rules eventually spread to monasteries across Europe, becoming a symbol of faith for devotees in a medieval era that otherwise ignored time. As something that set apart monks from the rest of society, the Benedictine system also engendered in laymen a sense that following a strict schedule of duties according to the clock was an important part of religious devotion. Eventually, the Benedictine's sense of time crept into everyday life and language. The word *siesta*, for instance, comes from the abbot setting aside an hour of rest after the midday meal at the sixth hour. Devout Catholics still pray at *matins* in the early morning and at *vespers* in the evening. Some historians believe that modern capitalism, with its use of time as an economic unit—for wages, contracts, and interest rates—grew in part out of the Benedictine fixation on measuring time.

When Cassiodorus was still a young man he met, and perhaps was taught by the third of our troika in Rome, an abbot named Dionysius Exiguus (c. 500–560)—"Little Dennis." Described as a Scythian— one of a barbarian people who a century earlier had been driven south by the Huns from their ancient home in the Caucasus—little is known about Dionysius other than his work on the calendar and on one of the first collections of official Catholic rules known as canons. He knew Boethius and Cassiodorus, but was probably older. Late in life Cassiodorus remembered him fondly as a brilliant scholar with a great fluency in translating Greek and Latin. Also an accomplished mathematician and astronomer, in 525—the year Boethius was executed—Pope John I (d. 526) asked him to calculate the Easter date for the next year. At the time this was part of an effort by the Roman church to wean itself from its sister church in the East, who long had treated the science of determining Easter like some arcane pharaonic secret, a mystery understood only by those steeped in the tradition of Aristarchus and Claudius Ptolemy. With a wave of his Latin quill Dionysius changed all of this, ending the long hegemony of Alexandria by co-opting their formulas and methods, freeing Rome at last from the time lords of this ancient city of stargazers.

Of course, Dionysius was careful to couch his work in terms that would be acceptable to the spiritualists of his day, insisting in explanations about his work that the holy day of Easter should be calculated "not so much from worldly knowledge, as from an inspiration through the Holy Spirit." He then promptly turned to astronomy and mathematics to make his calculations, adopting what in those days was the most accurate method available, the 19-year lunar cycle. Essentially he updated the table computed by the Alexandrian bishop Cyril, extending it for another 95 years, from 532 to 627.

We needn't plunge too deeply into the numeric complexities of these long-forgotten tables, although a brief dip will help explain what a man such as Dionysius knew and had to work with as he struggled to make sense of his Christian-Roman calendar. For example, in the chart below are four years in Dionysius's first 19-year cycle:*

Year (A.D.)	532	533	534	535
Indiction (I)	10	11	12	13
Moon's Phase (II)	0	11	22	3

*He is using the old Roman system of kalends, ides, and nones, which would linger throughout the Middle Ages.

Year (A.D.)	532	533	534	535
Day of the Week of March 24 (III)	4	5	6	7
Year in 19-Year Lunar Cycle (IV)	17	18	19	1
First Day of Passover (V)	Nones April	8 Kalends April	Ides April	4 Nones April
Easter Sunday (VI)	3 Ides April (April 11)	6 Kalends April (March 26)	16 Kalends May (April 16)	6 Ides April (April 8)

Below are explanations of each of the lines headed up by a Roman numeral:

I: This number has nothing to do with calculating Easter. It refers to a system of dating Roman documents in 15-year cycles called *indictions*, a style of dating so widely used for financial and legal documents (often in conjunction with the date of a consul or emperor's reign) that Dionysius included it as a helpful guide to the year for those using his table.

II: To calculate the true Easter, astronomers started by noting the "age"—or phase—of the moon during a given year on a set date in the solar calendar. This was arbitrarily set by Dionysius at March 22, the day after the official spring equinox as determined at the time of the Council of Nicaea. For instance, in 532 the moon's age was 0 days old on March 22—a new moon. This age-number is called an *epact*. Because the lunar year runs 11 days fast against the solar year, the age of the moon on any given date in the Julian calendar will always be 11 days "older" the next year. Thus in 533 the epact of the moon was not 0, but 11.

A year later, in 534, the epact moved another 11 days back, for a total of 22 days of movement since 532. But because the moon runs in a 29½ day cycle (rounded up to 30 days by Dionysius), the next year, 535, has an epact of 3, determined by taking $22 + 11 = 33 -$ the 30-day month $= 3$. And on it goes with eleven added to each year, running on a 30-day cycle.

The epact is important because in the 19-year lunar cycle this number will always be the same for each year in the cycle. (See number of the year in the 19-year progression.) This formula made it simple for anyone with even a rudimentary knowledge of numbers to calculate Easter, though later time reckoners would realize that the moon does not fit precisely into this cycle, since the lunar month is actually less than 30 days. Whether or not to use epacts became a hotly debated topic during the deliberations in the sixteenth century that led to the Gregorian reforms in 1582.

III: This is the day of week that fell on March 24, which was used to determine on which date the Sunday after the equinox would fall.

IV: The year in the 19-year lunar cycle.

V: This is the beginning of Passover, corresponding to Nisan 14 in the Jewish calendar—a date that Christian time reckoners were ordered to avoid by the bishops at Nicaea, who dictated that Easter could never be held on the day Passover begins. If the calculations for Easter indicate a date on Nisan 14, the celebration was moved to the following Sunday.

VI: The correct date each year for Easter Sunday, based on the formula in use at the time of the Council of Nicaea in 325. This has Easter falling on the first Sunday after the first full moon after the spring equinox.

Dionysius, like other Easter time reckoners past and present, provides numerous equations that prove the interconnectedness of these dates mathematically.* These are practical for the serious ecclesiastic task at hand but also seem in their elegance to be the product of a mind that enjoyed the precision and exactitude of equations for their own sake, despite his devout talk about "the Christian concept of time."

Dionysius's contribution to our calendar went far beyond the pedestrian task of calculating another 95 years of Easters. When he published his tables he included a reform that was little noticed in his own day but now affects virtually everyone in the world: the system of dating known as *anno Domini* (A.D.), "the year of our Lord"—which many people now call the *common era* (C.E.).

In a letter to a bishop named Petronius, Dionysius complained that earlier Easter tables used a calendar widely followed at the time, which started its year one in A.D. 284, the year that Emperor Diocletian ascended to the throne. Under this system, the year Dionysius wrote his letter—which we call A.D. 531—was designated the year 247 *anno Diocletiani*, the year of Diocletian. But Diocletian was a notorious persecutor of Christians, noted Dionysius, who tells Petronius that he "preferred to count and denote the years from the incarnation of our Lord, in order to make the foundation of our hope better known and the cause of the redemption of man more conspicuous." Dionysius calculated that Christ was born exactly 531 years earlier—which became *his* base year of A.D. 1. (Dionysius did not designate a year 0

---

*One flaw in Dionysius's system was the impossibility of matching up the seven-day week, in which Sunday fell, mathematically with a 95-year period of 19-year cycles. Obviously seven does not divide into 95, which meant this table was still not entirely accurate as a predictive tool. A mathematician in Aquitaine named Victorius figured out a solution to this problem c. 457 by figuring out that Easter dates repeat themselves every 532 years, 532 being a number divisible by 19 and by 7. Apparently Dionysius was unaware of Victorius's discovery.

because the concept of zero had not yet been invented). Where the abbot got this date for Christ's birth is unknown. Nor is it clear if his scheme was an original idea or one already informally used. Whatever the source, Dionysius was the first ever to use the system we all now take for granted when he wrote on his Easter tables *anni Domini nostri Jesu Christi* (the years of our Lord Jesus Christ) 532–627.

Unfortunately, Dionysius almost certainly got his dates wrong. The true moment of Christ's birth is unknown and a matter of immense controversy even today, given the vague and contradictory information available on Christ's early life. The Gospel of Matthew claims he was born in the time of Herod the Great, who died in 4 B.C. This means the birth must have occurred before this date. Other Gospels and historical sources suggest dates ranging from 6 or 7 B.C. to A.D. 7, though most historians lean toward 4 or 5 B.C. This means the year 1996 or 1997 was probably the true year 2000 in the *anno Domini* calendar, if one does the arithmetic without a year 0.

Anyway, it took time for Dionysius's use of *anno Domini* to catch on. Some Christians resisted it because they preferred the *anni Diocletiani*, also called the "Era of the Martyrs," a period held in veneration despite its association with an anti-Christian emperor. (Coptic Christians in Egypt still use *anni Diocletiani;* for them, the year A.D. 2000 will correspond to the year 1716 in the "Era of the Martyrs.") It was Dionysius's friend Cassiodorus who first used the A.D. system in a published work when he and his monks wrote their textbook in 562 on how to determine Easter and other dates, the *Computus paschalis.* Other Italians gradually accepted the A.D. system over the next several decades, followed very slowly by other regions of Christendom.

Early Catholic missionaries introduced the system in Britain, where newly converted Saxons issued edicts dated with *anno Domini* in the seventh century. It first appeared in Gaul during the eighth century but did not come into wide use in Europe until the tenth century. In some outlying provinces, including parts of Spain, the A.D. system was not adopted until the 1300s. Christians did not use the inverse of *anno Domini*, B.C. (for "before Christ") until 1627, when the French as-

tronomer Denis Petau apparently became the first ever to add B.C. to
dates while teaching at the Collège de Clermont in Paris.

Soon after the elderly Cassiodorus published his textbook on *computus*
in 562 the eastern emperor Justinian died, leaving his ambition to
reestablish the western empire unrealized. His efforts ultimately
proved disastrous to the West, as he and his immediate successors
found themselves overextended and unable to fend off fresh assaults
by Lombards, Bavarians, Saxons, and other Germanic tribes. Even
worse, these previously obscure invaders were far less Romanized than
the Germans Justinian had destroyed, *barbari* who had long associated
with Rome on the border of the old empire. With homelands deep
in the hinterlands of Europe, the newcomers were far more rapacious
and thorough in their ravaging and in establishing tribal-style governments. The Byzantines retained a toehold in Ravenna and in other
parts of Italy for several more decades, and remained a presence for
centuries to come. But in the wake of Justinian's juggernaut, most of
the West collapsed again into near anarchy, with the only remnant of
central authority residing in the Church.

Boethius's execution in 524 had signaled the instability of an age
that had little interest in intellectual pursuits. But the death of Cassiodorus sometime in the 580s—presumably safe behind his monastic
walls—symbolized the final gasp of an ordered world where time had
mattered and calendars framed how most people lived, worked, and
worshiped. With the West now a political and intellectual wasteland,
people had little need for formal civil calendars, with most reverting
back to a preliterate age when farmers, sailors, and merchants measured time as the Greeks did in Hesiod's days—in broad cycles where
events were triggered by the bloom of a flower or the flight north or
south of flocks of birds. For much of Rome's illiterate population this
had always been the way time was measured. But now, as Boethius

lamented in his *Consolation,* the entire culture seemed to be sliding into an abyss:

> For who gives in and turns his eye
> Back to the darkness from the sky,
> Loses while he looks below
> All that up with him may go.

Time had finally come to a full stop. Or at least it seemed this way, though remarkably a few monks and scholars over the coming centuries would keep the mechanism of calendar time moving, if barely. Indeed, the story of the calendar now shifts to one of the greatest of these medieval lights, a man who lived not in Rome or some other ancient center of culture, but on a shadowy island on the edge of what was to these Europeans the known world.

6
Monks Dream While Counting on Their Fingers

It is said that the confusion in those days was such that Easter was sometimes kept twice in one year.
—BEDE, A.D. 731

Under an ancient gnarled oak tree in southwest England the first archbishop of Canterbury held a meeting sometime in the late 590s—about a decade after Cassiodorus died in Italy—to settle a local dispute over Easter.

The archbishop, a Greek named Augustine,* was trying to convince a delegation of Celts from the western side of the island to abandon their system of calculating the Easter date, which deviated from St. Peter's. Isolated since the last imperial legion abandoned the island in 410, these Celts had been Christianized late in the Roman era only to find themselves cut loose soon after from both the empire and the Church in Rome. Since then waves of invasion by Saxons and Angles had driven these ancient Britons into what is now Wales, where they had joined with other Christian Celts from Ireland to form an independent church, with its own ideas about dating the Resurrection.

Augustine, dispatched to Britain by the pope to evangelize the Saxons and to Romanize the Celts, insisted that God was on his side. To

---

*This is not Augustine of Hippo.

95

prove it he reportedly performed a miracle under that old oak tree—restoring sight to a blind man.

The Celts were impressed but unconvinced. "Whereupon Augustine . . . is said to have answered with a threat that was also a prophecy," writes the British monk Bede (672–735), recounting the story a century later, "telling the Britons that their intransigence would one day cause their destruction."

Sure enough, wrote Bede, a few years later a brutal Saxon king named Aethelfrith (d. 616) "raised a great army and made a great slaughter of the faithless Britons." The dead included 1,200 unarmed monks massacred near their monastery at Bangor, south of modern Liverpool. That King Aethelfrith was a butcher intent on expanding his tiny kingdom at the expense of Celts; and that he was a pagan who cared nothing about Easter, hardly mattered to Bede and other Christians siding with Rome in this murky, little-known corner of Europe. For them the massacre was the fulfillment of Augustine's prophecy against these "faithless Britons, who had rejected the offer of eternal salvation, would incur the punishment of temporal destruction."

And what was the difference between the two churches' dates for Easter?

A single day.

You see, the Celts placed the date of Christ's crucifixion on a Thursday instead of a Friday. This meant their Easter had to fall (according to the Jewish calendar) between Nisan 14 and 20, while Rome said the date must fall between Nisan 15 and 21—a difference so minor that it is hard to imagine anyone quibbling to the point of bloodshed. Especially given the fact that Bede himself, one of the most brilliant time reckoners in the Middle Ages, knew something that almost no one else did in this murky era: that Rome's official dating of Easter was itself in error, because the Julian calendar it was based on was flawed.

Bede was almost sixty years old in 731 when he published his account of the prophecy and slaughter in his *Ecclesiastical History of the English People*. A monk, teacher, and choirmaster at the Saxon-era monasteries of Wearmouth and Jarrow in Northumbria, he lived far away from the centers of culture and learning (such as they were) in his age— which makes his accomplishments all the more astonishing. For without ever leaving the neighborhood of his twin monasteries, Bede wrote some sixty books on subjects ranging from commentaries on the Bible to works on geography, history, mathematics, and the calendar. He penned detailed letters describing the concept of the leap year, his calculations about the supposed motion of the sun around the earth, and his measurements of equinoxes. He even came up with the name *calculator* to describe a time reckoner, and later *catholicus calculator*—"Catholic calculator."

"I was born on the lands of this monastery," Bede wrote in his *History*. "I have spent all the remainder of my life in this monastery and devoted myself entirely to the study of Scriptures. And while I have observed the regular discipline and sung the choir offices daily in church, my chief delight has always been in study, teaching, and writing." Handed over to the abbot of the monastery by his apparently upper-class family at age 7, he was educated by the monks, became a church deacon at age 19, and was ordained a priest at age 30—all at Wearmouth and Jarrow.

Built in the latter part of the seventh century, Wearmouth was founded shortly after Bede's birth in 672 on the coast of England near where the River Wear pours into the North Sea—a country of rolling hills, limestone and sandstone outcrops, low mountains, and ruined Roman walls and towns. The monks built a companion monastery nine or ten years later at Jarrow, a few miles away on the mudflats at the confluence of the Don and Tyne rivers. Both began as Saxon structures of timber and straw until one of the project's sponsors, a monk of noble birth named Benedict Biscop (c. 628–690), decided the buildings should look like the stone churches he had seen during his travels in Gaul. With Hadrian's ruined wall and an old Roman fort nearby, stone was readily available for pilfering, though Benedict Biscop had to bring over skilled labor from Gaul because Britain lacked master builders and stone masons. He also brought across the channel glassmakers who glazed the windows and made glass receptacles.

Benedict filled his buildings with a rich assortment of imported altar vessels, paintings and carvings—and with a library. Taking five trips to Rome Benedict brought back "a great mass of . . . books," including calendars—among them almost certainly Dionysius Exiguus's charts and calculations, and the latest martyrologies (lists of saint's days and other holy dates). The exact contents of Benedict's library is unknown, though it seems to have included a copy of a Bible used and illustrated by Cassiodorus, known as the *Codex Grandior*, as well as theological works, a smattering of Greek philosophy and mathematics, and Cassiodorus's encyclopedias of ancient knowledge.

It was an impressive library for its day, though at best it contained some four to five hundred works.* This compares to perhaps two to three *thousand* volumes Cassiodorus had access to a century and a half earlier in his library, which itself was profoundly diminished from the vast collections of antiquity, including Alexandria's library and its four hundred thousand manuscripts. Imagine what this meant to the inquiries of the second-century astronomer Claudius Ptolemy, who had a mountain of information at his disposal, compared to Bede. Working six hundred years later in his cold monastic cell at Jarrow, Bede had to make do with just a few treasured vellum scrolls tucked into wooden boxes to keep them from rotting in the dampness common to Northumbria.

Likewise, Bede and his countrymen were only vaguely aware of events beyond the frigid, turbulent waters of the *Mare Germanicum,* now known as the North Sea. It probably took several years, for instance, for Northumbrians to find out that the mother church in Rome had finally broken off its titular allegiance to Constantinople,

---

*Bede cites about 175 sources in his writings.

which had claimed authority over the former imperial provinces of the West as the inheritor to Rome—a claim that had become increasingly unrealistic after the failed attempt of Justinian to reconquer the West. In part this break came about because of another seismic event happening far from the British Isles—the sudden appearance of Islam in the mid-seventh century, which eventually forced the Byzantines to recall their legions from central Italy. Following Mohammed's teaching and his founding of the first mosque at Medina in 622—year 1 in the Moslem calendar—the armies of Islam had swept like a firestorm to seize Arabia, Mesopotamia, Persia, and Egypt by 651; North Africa by 702; and Spain and parts of Asia Minor by 711, when Bede was about 38 years old. By then the stunned Byzantines had lost nearly their entire empire, and were fortunate to have held on to their heartland in western Asia Minor, coastal Greece, and Sicily.

Meanwhile the politics in the West remained confused, with shifting tribes battling, conquering, and being conquered. Lombards reigned for the moment in northern Italy. East of the Danube lived pagan Slavs, who had gradually enveloped much of the former provinces of Rome in northern Greece and in the Balkans. Closer to Britain, the Franks had dominated what is now France and Germany for over a century; in 732, a year after Bede published his *History,* the Merovingian kings of France decisively beat back the Moslem invaders of Spain as they attempted to roll into southern France.

In faraway Britain this was at best a distant rumble, though it's likely that Bede himself felt far more isolated intellectually than geographically. Indeed, he lived in a time when even monks in monasteries were turning away from all but a crude understanding of basic scholarship, either because they lacked manuscripts and teachers or because they had no use for knowledge they considered ungodly and profane. Most aspired to follow Cassiodorus's admonition to learn, though few succeeded beyond a clumsy understanding of basic concepts. In France one senior cleric complained that many monks and churchmen were completely illiterate. At Jarrow Bede himself had to translate the

Lord's Prayer from Latin to the local vernacular so that his brothers could understand the Latin words they spoke when they prayed.

Scholarship in many places was reduced to learning a few key subjects by rote and devoting one's life to copying ancient manuscripts, which most monks held in awe as artifacts of a glorious past, but few understood. A number of monks lost their eyesight scratching out copies in the semidarkness of their stone cells, since candles were not allowed for fear fire would consume the ancient parchments. "He who does not turn up the earth with the plough," a sixth-century monk admonished his brothers, "ought to write parchments with his fingers." Many monks did not stop with mere writing, but also adorned their manuscripts with stunningly beautiful ornaments, calligraphy, and illustrations: glittering gold-leafed letters and painted flowers and vines; masterly images of winged angels, fiery demons, tortured saints, and Christ enthroned in heaven. Some of the most dazzling illuminations appear on medieval calendars, which typically list month-by-month dates and saint's days and are lavishly illustrated with scenes of peasants gathering hay in June, nobles hunting and drinking wine in August, and peasants huddling beside hearth fires as snow blankets the out-of-doors in February.

If few of these monks thought deeply about the knowledge in these lovely books, fewer still came up with their own interpretations about time reckoning or anything else in the scientific realm. This makes a genuine scholar such as Bede all the rarer. In fact, the only other truly notable time reckoner in these dark days of the early Middle Ages was Isidore of Seville (560–636), a Roman ecclesiastic and scholar living in another distant outpost of the former empire: Visigoth Spain. The Archbishop of Seville, Isidore is known for eradicating Arianism among the Visigoths and stifling other so-called heresies in Spain— and for compiling a great encyclopedia along the lines of Cassiodorus's, a *summa* of universal knowledge as it existed in this sunny, hot corner of Europe. Preserving numerous fragments of classical works that otherwise would have been lost, he described the fundamentals of general astronomy and mathematics, including a section on time reckoning and the Easter cycle that would be used by Bede and other time reckoners over the next few centuries.

Yet even Isidore's work follows the tendency of this era to substitute copying and the reiteration of past thinking for true scholarship. Little in his encyclopedia is original, and some of it is poorly written. Isidore even apes Cassiodorus's admonition to learn and understand astronomy and mathematics, offering little analysis or insight of his own. "Remove *computus* from the world," Isidore wrote, essentially plagiarizing an almost identical statement made by Cassiodorus, "and everything is given over to blind ignorance. . . . If you remove the number from objects, then everything collapses."

This encouraged many a medieval monk to embrace the science of *computus,* though at the same time Isidore, like Cassiodorus, instructed his brothers to think of timekeeping devices as mere tools, like a key or a chain—an admonition that reinforced the medieval tendency to rely on already established equations and rules that required little imagination or creativity, a process that perpetuated the prevailing simplification of Augustine of Hippo's view that understanding time beyond a simple calendar and dating Easter was better left to God.

During this period most of Europe still followed Julius Caesar's basic calendar, though pagans beyond the Christian realms continued to use their own ancient calendars. To the north the Saxons (those who had not emigrated to Britain) and other old German tribes used a combination lunar-solar calendar that started with the twelve lunar months and then added a month every so often to match it up with the solar year. This calendar began on December 25, shortly after the winter solstice. Month names included the third month, Solmonath, the month of offering cakes; Blodmonath, the month of sacrifice; and Eosturmonath, named after the goddess of spring and twilight, Eostre. Another modern word derived from the Saxon calendar comes from

Guili, the name of the Saxons' first and last months of the year. The Old English for Guili is *geol;* in modern English it is *yule.* Guili occurred during winter, hence "yule log" and "yule season."

The Slavs who dominated eastern Europe during Bede's day apparently used a purely lunar calendar. Islam, symbolized by the crescent moon, also ignored the sun and still does in its religious calendar, which drifts across the solar year at a pace of eleven days a year. Father east, the Chinese under the T'angs—one of the richest and most stable dynasties in Chinese history, then at the height of its power and influence—continued to use a calendar similar to those developed in ancient Babylon and Greece a thousand years earlier. Based on a lunar year, this calendar added extra months seven times during a 19-year cycle. They assigned numbers to identify each month, but used their zodiac symbols to name years in a 12-year cycle of animals familiar to anyone who has eaten in a Chinese-American restaurant with a printed placemat listing the year of the rat, ox, tiger, hare, dragon, snake, horse, sheep, monkey, rooster, dog, and pig. The Gregorian year 2000, for instance, is the year of the dragon. The year of Bede's death in 735 was the year of the pig.

From the perspective of a T'ang astronomer in 735 it would have been laughable to imagine that Bede's calendar would one day become the world's. Still, even as invaders on all sides conquered territories once Christian, the seeds were being sown to expand the hold of Christianity—and by default the Julian calendar. Christianity had always been a proselytizing religion, taking literally the words of Christ when he said, "Follow me." Like Islam, it offered a potent and coherent set of religious ideals and duties that proved highly attractive to religiously minded people. Also like Islam, it had fused its doctrines and faith with the apparatus of political power—first under the aegis of Rome and more recently under the sponsorship of barbarian kings converted to Christianity. This made the spread of Christianity less an individual decision than a strategic ideology of kings, nobles, and through them entire peoples.

By the time Bede was a young man, the Church's conversion of *barbari* and the conquests of Islam had precipitated a titanic shift in Christianity's geography, transforming it from a religion primarily of the Mediterranean and Near East to a European religion. The most critical moment had come sometimes between 496 and 506 when King Clovis of the Franks agreed to be baptized by a Catholic bishop at Reims. A shrewd politician, Clovis embraced Rome to gain the support of Gallo-Roman Catholics in his successful war against the Arian Goths in what is now central and northern France. Clovis's victories set in motion a kingdom that would eventually split into France and Germany, nations that for centuries remained closely connected with the Church in Rome. The Catholics made further inroads with other Germanic tribes, though Christians in Bede's day were hardly of one mind. The Goths, Burgundians, and Alemanni remained Arian, which was only one of several sects that deviated in ways large and small from official Roman doctrine. Arians, for instance, continued to worship Easter according to their own formulation of dates, as did a remnant of the Celts whose brothers had been massacred by Aethelfrith a century earlier.

Most of the Christian expansion into Germanic countries remains murky. Details are recorded, if at all, by scattered letters from bishops and popes in Rome and by local chroniclers of Franks and others whose grammar and grasp of literary style was poor and their facts jumbled or suspect. England is an exception because of Bede. But his *History* is important beyond the stories it tells because Bede chose to use Dionysius Exiguus's scheme of *anno Domini* to date the events in his chronology—the first time this was done in such a prominent and widely read history. He also agreed with Dionysius's dating of Christ's birth, affirming the Scythian monk's designation of the year 1 that we still use today. Before Bede, historians had dated events using the reigns of kings and emperors. Or, like the ancient Greek historian

Herodotus, they had simply strung together stories roughly in chronological order with no precision in exactly when they took place.

Bede's history starts with brief sketches describing the island and original inhabitants of Britain, its conquest, rule, and abandonment by Rome, its invasion by Saxons and Angles in their long boats, and the two centuries of chaos that followed as the Germans fought among themselves and against the old Romano-British population. Bede then settles into the meat of his story when Rome in the time of Archbishop Augustine looked once more toward Britain—as a country to conquer not militarily but spiritually. Anyway, it seemed to be the next logical step for expanding the Christian reach once Gaul was firmly in the Catholic sphere. Yet Bede insists that the pope who dispatched Augustine to Britain in 596, Pope Gregory I (540–604), was inspired less by strategy than by compassion. Bede tells the story in his *History:*

> We are told that one day some merchants who had recently arrived in Rome displayed their many wares in the market-place. Among the crowd who thronged to buy was Gregory, who saw among the merchandise some boys exposed for sale. These had fair complexions, fine-cut features, and beautiful hair. Looking at them with interest, he inquired from what country and what part of the world they came. "They come from the island of Britain," he was told, "where all the people have this appearance." He asked whether they were Christians, or whether they were still ignorant heathens. "They are pagans," he was informed. "Alas!" said Gregory with a heartfelt sigh: "how sad that such bright-faced folk are still in the grasp of the author of darkness."

Gregory asked the name of the slave boys' race and was told they were Angles. "That is appropriate," he said, "for they have angelic faces, and it is right that they should become joint-heirs with the angels in heaven."*

---

*This loses something in translation. In Latin Gregory said: *"Non Anglii, sed angeli,"* literally: "Not Angles but angels."

Whether moved by the boy slaves or by politics, Pope Gregory in 596 had dispatched Augustine, a Greek monk and Gregory's former monastic roommate, from St. Peter's to evangelize the distant Britons. It says a great deal about the state of Europe's highways—and the immense distance to Britain in the mind-set of these Romans—that when Augustine and an entourage of forty monks "progressed a short distance on their journey, they became afraid, and began to consider returning home. For they were appalled at the idea of going to a barbarous, fierce, and pagan nation, of whose very language they were ignorant." The monks became so fearful that they voted to send Augustine back to Rome "so that he might humbly request the holy Gregory to recall them from so dangerous, arduous, and uncertain a journey." Gregory understood their reluctance but ordered them to continue. This reply came in a letter from Gregory that demonstrates the dating system then in use—one that had not yet incorporated Dionysius Exiguus's new *anno Domini* concept. After exhorting the monks to continue on to Britain, and asking that "God keep you safe, my dearest sons," Gregory recorded the day he wrote his letter:

> Dated the twenty-third of July, in the fourteenth year of the reign of the most pious Emperor Maurice Tiberius Augustus, and the thirteenth year after his Consulship: the fourteenth indiction.

The emperor referred to is Maurice of Constantinople, whom the Romans at this time still nominally regarded as the titular ruler of the West; the "indiction" is the year in the fifteen-year cycle that had been used since Diocletian's time to date Rome's financial and legal dealings.

During the days of the empire the journey north from Italy through southern France, and onward to Britain took several days through a settled country over good roads. In 596 the journey took weeks to pass through territories thick with thieves, marauders, and stretches of land once peaceful and under till, but now abandoned. Traveling by ship across the channel, Augustine arrived at Ebbsfleet on the island

of Thanet, where the Germanic king of Kent, Aethelberht, met him
in the open air. Aethelberht was married to a Christian princess from
the Frankish royal house but remained a pagan himself. He chose an
open field, Bede says, because "he held an ancient superstition that,
if they were practisers of magical arts, they might have opportunity
to deceive and master him" should he meet them in a more enclosed
space. Arriving in full regalia and carrying a cross of silver and a picture
of Christ, Augustine and priests made a favorable impression on the
king. He even provided them an old basilica in his capital at Canterbury that long ago had been a Christian church under the Romano-
Britons—a move that Bede says paved the way for Aethelberht to
convert by 601, when Augustine was consecrated archbishop of Canterbury.

With the conversion of the Saxons came the reintroduction of Caesar's calendar in Britain, with certain Anglo-Saxon modifications. For
instance, the substitution of Germanic planet-gods for those of Rome
to designate the days of the week, and the use of the goddess Eostre
to name Easter—which then and now is officially called the Feast of
the Passion by Catholics. This followed an already long tradition in
the Church of absorbing certain pagan customs into local ceremonies
and beliefs. This policy was spelled out by Pope Gregory in another
letter, where he tells Augustine not to work the Saxons' pagan temples:

> The idols are to be destroyed, but the temples themselves are to be aspersed
> with holy water, altars set up in them, and relics deposited there. . . . In this
> way, we hope that the people, seeing that their temples are not destroyed,
> may abandon their error and, flocking more readily to their accustomed
> resorts, may come to know and adore the true God. And since they have
> a custom of sacrificing many oxen to demons, let some other solemnity be

substituted in its place, such as a day of Dedication or the Festivals of the holy martyrs whose relics are enshrined there. . . . For it is certainly impossible to eradicate all errors from obstinate minds in one stroke. . . .

Dated the seventeenth of June, in the nineteenth year of the reign of our most pious Lord and Emperor Maurice Tiberius Augustus, and the eighteenth after his Consulship: the fourth indiction.

Gregory does not specifically mention days of the week or the Saxon naming of Easter as part of his campaign. But it is not too much of a stretch to assume that Tiw's day, Woden's day, Thor's day, and Freya's day—and Easter—came to be used in early Christian England as part of an effort to win over the "obstinate minds" of Saxons and Angles.

When Augustine arrived in Britain in 597 he was, at best, only vaguely aware that Christians already lived on the island—the Celts he would soon meet under the old oak tree. Indeed, these Celts and Romano-Britons may have lost ground against the Germans, but they had been gaining ground for their Celtic church as they proselytized across Ireland, Scotland, and northern England, winning souls among the Celtic pagans—who adopted the Celtic system of dating Easter—even as Augustine showed up in the south and began evangelizing for the Roman church.

Both sects built large monasteries and competed for converts, with Northumbria becoming a major spiritual battleground during the time of King Oswiu (612–670), who embraced the Celtic faith. Then he married the Princess Eanfled of Kent, a Catholic, who brought with her from Canterbury her own bishop and priests. This introduced two dating schemes for Easter to the royal court—which in most years did not matter, since the Celtic and Catholic calculations were not far off from each other. But every so often—such as in 664—the dates differed. "It is said that the confusion in those days," writes Bede, "was such that Easter was sometimes kept twice in one year, so that when the King had ended Lent and was keeping Easter, the Queen and her attendants were still fasting and keeping Palm Sunday." For Christians this was horrific: the royal couple, representing law and truth for their subjects, celebrating the holiest day in the kingdom on separate dates. For people of this period the discrepancy went far beyond a religious squabble. It undermined the order of the state—such as it existed in this still murky time—and of a universe that was supposed to provide absolute answers from an infallible God.

At least this was the theory. In reality, the king and queen's rival Easters were tolerated for several years—until Oswiu's son, trained by the Catholics, convinced his father that something should be done if the country was to have the sort of unified church rulers sought in those days. So in 664 defenders of both traditions gathered for a conference to decide the issue at the monastery of Streanaeshalch—the "Bay of the Beacon"—at Whitby, on the coast some 40 miles north of York. Bede tells us it was a cordial, if sometimes passionate, exchange, an outback version of Nicaea in 325, where rival sects gathered to feast and freely debate before a sovereign who at the end would make a decision affecting the future of the holy days.

An Irish bishop named Colman argued the case for the Celts, invoking the authority of the apostle John to defend his church's dates. On the Roman side an abbot named Wilfred cited the authority of Nicaea and other councils, adding that "a few men in a corner of a remote island should not be preferred before the universal Church of Christ throughout the world." Would these scruffy islanders, Wilfred asked, remain backward and outside the mainstream of European culture, or would they join the same mighty Church championed by the Franks and other kingdoms?

King Oswiu was no fool. He believed in the Irish teaching he had grown up with, but he also understood that it made little sense to remain obstinate against Rome and the rest of Catholic Europe. So in the end he decided to abolish the Celtic system and to adopt the Roman, saying he was especially swayed by Wilfred's argument that the pope, as the successor to St. Peter, had the authority to decide Church dogma. Wilfred quoted Christ as saying: "Thou art Peter, and upon this rock I will build my Church." More to the point for the literalists of the early Middle Ages, Wilfred quoted Christ as saying he gave to Peter "the keys of the kingdom of heaven." Oswiu responded by asking the Celtic bishop if Christ actually said these words. Bishop Colman admitted he did, and that the Celts had no such authority given to the founders of their church. Bede, a Catholic himself, tells us what the king said next:

> "Then, I tell you, Peter is the guardian of the gates of heaven, and I shall not contradict him. I shall obey his commands in everything to the best of my knowledge and ability; otherwise, when I come to the gates of heaven, there may be no one to open them, because he who holds the keys has turned away."
>
> When the king said this, all present, both high and low, signified their agreement and, abandoning their imperfect customs, hastened to adopt those which they had learned to be better.

This was not entirely true. Several Irish hard-liners returned to their bleak monastery on the Scottish island of Iona and continued to flout Rome. These included Bishop Colman, who retreated first to the Celtic monastery on Iona and then to western Ireland with thirty monks to avoid accepting the Roman calculation of Easter. As late as 687, a quarter century after Whitby, the Irish-trained bishop Cuthbert admonished die-hard Celts to stay the course with Rome. He told his disciples to have "no dealings with those who had wandered from the unity of the Church either through not celebrating Easter at the usual time or through evil living."

Soon after the synod the pope dispatched Theodore of Tarsus, a native of Asia Minor, to take over as archbishop of Canterbury. His name had been at the bottom of the short list; he was apparently selected because several others refused to take the position. It was under Theodore that the monasteries at Jarrow and Wearmouth were founded by Benedict Biscop. Theodore also oversaw the religious integration of the Celts and Catholics leading up to Bede's time, a period that turned out to be a brief moment of near stability in Anglo-Saxon Britain. It lasted until the ninth century, when the first Viking long boats appeared off the beaches of Northumbria.

Bede in his *History* is clearly a partisan of the Catholics' method for dating Easter. But he does not leave it at that. As a teacher and practitioner of *computus,* he set out to prove that the Church was correct beyond a doubt about the true Easter. This effort began modestly in 703, when he was about thirty years old. He wrote a short work on time reckoning, *Liber de temporibus,* for his students: a combination how-to, analysis, and refutation of the Christian Celts' stand on Easter. In this work the young Bede also confirmed Dionysius's system of 19-year cycles to determine Easter and his use of *anno Domini.* This seal of approval brought these systems into the mainstream of the Middle Ages, which widely read and revered Bede over the next several centuries. In 725 he wrote a longer version of the *Liber de temporibus* at the request of his students, titled *De temporum ratione,* a tome that has been found in over a hundred libraries and collections of medieval manuscripts across Europe, attesting to its popularity. No comparable scientific work was written about time and the calendar in the Latin world until the era of Roger Bacon, almost five centuries later.

De temporum ratione and the shorter pamphlet are part compilation of known ideas and part original thinking. Bede started with an assumption that might have made Augustine squirm: that the universe as created by God was a place of order in which all phenomena could be rationally and logically explained, even if much of it was beyond human comprehension. Following the ancients, he writes that this universe consists of the elements earth, air, fire, and water, and that the earth lies at the center—surrounded, as Christian theology taught in that period, by seven heavens: air, ether, Olympus, fiery space, firmament, the heaven of angels, and the heaven of God. (This is where we get the term "seventh heaven" to describe something truly wonderful.) He provides primers on how to count to one million using one's fingers—the only handy counting device available to Bede—and how to master Roman and Greek numerals. He also explains the divisions of time as they then existed, following Isidore of Seville's list, from the smallest unit to the largest: moments, hours, days, months, years, centuries, and ages.

Bede also writes about the long-held Christian belief that the earth had passed through six ages since the Creation. The first five, he said, had been marked by the Creation, the Flood, Abraham, David, and the captivity of the Jews in Babylon. The sixth and current age began with the birth of Christ. This idea of a "calendar" of six ages came from the words of the apostle Peter. He says in the Bible that "one day is with the Lord as a thousand years, and a thousand years is one day." In the Middle Ages Christian chronographers interpreted this to mean that each age of the earth would last roughly a thousand years. This was probably not Peter's intent, since he seems to be saying in this passage that time to God is meaningless because he is omnipotent and timeless. Nevertheless, Western chronographers before and after Bede used this passage to date the beginning of the world to about five thousand years before Christ's birth.

Bede, however, studied the problem and came up with his own dating of the five ages, based on a careful reading of Old Testament texts translated directly from Hebrew to Latin rather than relying on third- or fourth-hand translations from Hebrew to Greek to Latin. He concluded that the time span from the Creation to the birth of Jesus was 3,952 years. As for the duration of the sixth age—after which Christ himself was supposed to inaugurate a seventh and final age of heaven on earth—Bede stuck with Augustine of Hippo's admonition to avoid trying to predict the future, which the monk from Jarrow agreed that only God knows.

Incredibly, Bede's calculation of the first five ages of the earth led to an accusation of heresy, because his time span was at odds with those of other revered chronographers, including Isidore. Someone at a Saxon feast held at Jarrow shouted the allegation after a liberal amount of alcohol had been consumed. The charge infuriated Bede; he shot off a letter defending himself that suggested his accusers were ignorant fools. Apparently nothing came of the accusation.

In the sections of *De temporum ratione* about Easter, Bede calculated the holy day up to the year 1063 using Dionysius Exiguus's basic system of calculations, with one change. Instead of figuring the dates in arbitrary 95-year periods Bede used a 532-year cycle in which the Easter date repeats itself, based on multiplying the 19-year lunar-solar cycle times four (to account for the leap year) times seven (the cycle of a week from Sunday to Sunday). At least one earlier mathematician had stumbled on this cycle, though Bede was the first to use it systematically.

But Bede was not content simply to record categories and make calculations like other computists before and after him. Turning to empirical observation, he designed a complicated sundial that he checked every day to keep track of equinoxes. He hoped this would provide him with an objective estimation of the true Easter. In 730 he set out to prove to a friend that the equinox did not fall on March 25, as some insisted. Bede confirmed this with his sundial and kept up his daily record of the shadows cast to show that another equinox fell on September 19, 182 days later. Continuing his observations for another six months, he discovered that the spring equinox in 731 did not fall on precisely the same line *(horologii linea)* on his sundial as before, suggesting that the leap-year system of 365¼ days was not entirely accurate. This was an extraordinary find for a man using a sundial in dark-ages England. It is a pity that he had no working knowledge of a more accurate timepiece, such as a water clock. As it was, Bede had no way to divide the solar year into units smaller than very basic fractions, which means he had no way to quantify his discovery. He also got the true spring equinox wrong, since by 731 the error in the Julian calendar had caused it to drift more than six days since Caesar's reform in 45 B.C. This put the true spring equinox during Bede's experiment at March 18 or 19.

Bede, a refreshingly candid critic of his own work, suspected that his calculations were not entirely accurate. He invited others to improve them while he kept working to refine his observations himself. Since northern Britain is far less sunny than the Mediterranean, and lines of shadow on even the best sundial face are fuzzy and lacking in detail for many months of the year, Bede looked around for other natural time markers. He discovered tides. Taking long walks along the sand-and-rock coast of Northumbria, he seems to have kept a close, scientific eye on the ebb and flow of the ocean, eventually figuring out how to use the tides to measure the phases and orbit of the moon. He used them to concoct a formula for finding the zodiac sign the moon was passing through given its phase, allowing him to come up with an improved method for fixing the age of the moon on the first day of a given month. This project had little to do with the Easter computation, but it proved useful for astrologers, who used his zodiac equations to predict the future in a way that would have greatly disturbed the pious monk from Jarrow.

Bede went further than most toward embracing objective science, but he remained limited by the mind-set of his era's spirituality. We cannot forget that Bede was primarily a religious man devoted to his canonical duties and that most of his scholarly work was not scientific, but religious. We must also remember that Bede counted with his fingers as much out of choice as necessity, reiterating the familiar explanation that monks were not supposed to delve too deeply into the details of God's creation. When confronted with the need for complex fractions in calculating time, he simply rounded up or down, perhaps insisting like Isidore that God's reckoning, so far as it was understood, consisted of single-digit numbers—the ones that could be counted on the fingers of a human hand. Likewise he taught that there was no need to measure half or quarter hours with water clocks; that the "God-given hour" was a small enough unit of time. He told his students to use the twenty-four-hour system for scholarly purposes, but warned them this had no application to everyday life, particularly for the *vulgus* (great masses), who had no way to measure hours accurately and seemed to prefer the informal system of "hours" gauged by looking up at the position of the sun. "It is not for man to know the moments set by God," Bede quoted from the Bible.

Still, Bede devised a clever theory that attempted to explain the apparent discrepancies between secular time and sacred time. He suggested that there exist three categories of time—time determined by nature, such as the solar year of 365¼ days; time fixed by custom, such as the 30- and 31-day months that belong to neither the solar year nor a lunar phase; and time set by an authority either human or divine, such as the Olympiad every four years or the Sabbath every seventh day. Like Augustine of Hippo, he believed God's time superseded all other forms. This hierarchy of truths about time allowed Bede to embrace science, on one hand, and a world of miracles and God's omnipotence, on the other. Indeed, this is a man who carefully studied sundials but also filled his histories with stories about bishops curing blind men and prophecies about monks being massacred.

Bede finished his *Ecclesiastic History of the English People* four years before his death in 735, concluding with an enigmatic statement. "What the result of this will be the future will show," he writes, a curiously modern-sounding mix of pride and uncertainty, and a sense that humanity still had more to learn—a notion rare in his era, when most people believed that mankind had attained all of the knowledge it ever would, and that the world would soon end. Remembered fondly by his own peers and by subsequent generations, Bede's furtive embrace of the scientific method was centuries before his time, and would later amaze and hearten scientifically oriented thinkers such as Roger Bacon.

Even more remarkable was Bede's appreciation of time as being real and measurable; something that could be organized into an ordered system of epochs, years, months, and days. For Bede time moved in a progression along a calendar, a concept few people in his day embraced, as they lived from season to season and endlessly passed the hours, whether in sowing or chanting designated Psalms at the appointed time each day. This was perhaps Bede's greatest achievement: that he almost single-handedly kept time moving when everywhere else it had stopped.

7

Charlemagne's Sandglass

Time belonged only to God and could only be lived out. To grasp it, measure it, or turn it to account or advantage was a sin. To misappropriate part of it was theft.

—JACQUES LE GOFF

Sometime around the turn of the ninth century the first Holy Roman Emperor, Charlemagne (742–814), was said to have acquired a sandglass large enough that it ran a full twelve hours before it needed to be turned.* Details of what this timepiece looked like are not recorded. One imagines teams of strong men in Frankish costumes— tights, loose tunics, and bands of cloth wrapped around their legs— standing ready to flip a giant contraption made of polished wood and blown glass, filled with hundreds of pounds of sand. Looking on was an emperor in late middle age who by then had inherited or conquered virtually all of modern France, the Spanish Pyrenees, Belgium, the Netherlands, Germany, Austria, Luxembourg, Switzerland, Corsica, northern and central Italy, and parts of the Czech Republic and the Balkans. It had been four centuries since so much territory in western Europe was unified under the rule of a single man.

Charlemagne, with a flowing beard, protruding belly, and large,

---

*Sources are unclear about whether or not this sandglass existed. Most accounts do not mention it at all, with some experts contending that the sandglass was not invented until much later, in the thirteenth or fourteenth centuries. Others say sandglasses existed as early as the second century, B.C.

animated eyes, was a ruthless warrior who spent most of his seventy years in the saddle leading countless campaigns. He loved to eat game roasted on a spit, ignoring his doctor's warnings that it was bad for his health. At night he listened to storytellers recount Frankish legends and excerpts from Augustine's *City of God*. He also was fascinated by timepieces. Besides his twelve-hour sandglass, he received in 807 a famous gift from Sultan Harun ar-Rashid (766–809), fifth caliph of the Abbasid Dynasty and master of the Islamic world.

Best known to Eurocentric Westerners as the sultan in *The Thousand and One Nights,* al-Rashid's reign in Baghdad is known as a golden age for art and science in the Arab world, a period when the conquerors who had burst out of Arabia a century and a half earlier were settling down and integrating Islamic, Hellenistic, Persian, and Indian cultures under their rule. They created a great flowering of learning of the sort that Charlemagne could only dream about in his cold stone-and-timber castle at Aachen, his capital west of the Rhine in what was then a landscape of rolling hills and dense forests near modern Bonn.

Responding to an embassy sent by Charlemagne, the caliph dispatched to Aachen a number of gifts: an elephant, a luxurious Persian tent, silk robes, perfumes, ointments—and an elaborate clock. It was made of brass, "a marvelous mechanical contraption, in which the course of the twelve hours moved according to a water clock, with as many brazen little balls, which fell down on the hour and through their fall made a cymbal ring underneath. On this clock there were also twelve horsemen who at the end of each hour stepped out of twelve windows, closing the previously open windows by their movements."

For Charlemagne, such timepieces represented learning and progress, much like a Model T or an early Remington typewriter once signaled modernity in small, isolated towns across America. But al-Rashid's gift also must have underscored the Europeans' backwardness. They had nothing approaching such a wondrous device as the caliph's clock, a situation Charlemagne reportedly understood and deplored. Indeed, this remarkable warrior, when not off conquering, devoted considerable energy during his 47-year reign to support learning and a respect for intellectual pursuits notably lacking since the dismemberment of Rome four centuries earlier. Encouraging literary scholarship, architecture, and art, Charlemagne issued decrees requiring all priests to be well versed in basic knowledge. "Let those who can, teach," he ordered in 789.

He also insisted that his subjects learn and teach *computus* after hearing that few bishops or priests understood enough about mathematics and time reckoning to make competent calculations for the Easter holidays, or to maintain the Christian calendar. "Let the ministers of God's altar . . . collect and associate with themselves children . . . that there may be schools for reading-boys," Charlemagne commanded in his 789 edict. "Let them learn psalms, notes, chants, the *computus* and grammar, in every monastery and bishop's house."

Following the lead of Caesar and Constantine, who transformed their calendars as part of grand schemes to launch new political and religious eras, Charlemagne attempted to reform his calendar, too. Most important, he and his scribes incorporated into the civil machinery of his empire the *anno Domini* system of dating favored by Dionysius and Bede. Charlemagne also followed in many of his decrees a growing trend in Europe to number the days of the months in sequential order instead of using the cumbersome Roman system of kalends, nones, and ides. On Charlemagne's tomb, planted in the center of the octagonal cathedral he built at Aachen, the inscription reads:

> In this tomb lies the body of Charles, the Great and Orthodox Emperor, who gloriously extended the kingdom of the Franks, and reigned prosperously for forty-seven years. He died at the age of seventy, the year of our Lord 814, the 7th Indiction, on the 28th day of January.

The emperor also tried to Frankify the names of the months, with less success. He proposed naming the months after the seasons of the year, festivals, and holy celebrations. Under Charlemagne's system, January became Wintarmanoth, meaning "the month of cold," and April became Ostarmanoth, still another reference to the goddess Eostre or Ostar, namesake for Easter. Though it never caught on, this calendar did have far more relevance to Franks in the late eighth and early ninth centuries than months designated by Latin tribes on the Tiber a millennium and a half earlier, who named their lunar months after goats, pagan gods, and Latin numbers. Charlemagne's months run as follows:

Charlemagne's Months	Roman Months
Wintarmanoth	January
Hornung	February
Lentzinmanoth	March
Ostarmanoth	April
Winnemanoth	May
Brachmanoth	June
Heuvimanoth	July
Aranmanoth	August
Witumanoth	September
Windumemanoth	October
Herbistmanoth	November
Heilagmanoth	December

In the midst of the darkness enshrouding Europe this sudden passion for an intellectual life seems a miraculous turnaround. Here was a barbarian king, disgusted with the low ebb of learning, throwing open his court to what his own chroniclers describe as a virtual cult of scholarship. At Aachen and elsewhere Charlemagne's scholars, artists, and musicians collected manuscripts, published histories and ballads, and corrected translations of the Bible. His architects and engineers built a 500-foot-long bridge over the Rhine at Mainz and erected numerous churches and palaces, including the magnificent Aachen Cathedral, a classic of the Romanesque-Byzantine style. Famous for its wide arches and octagonal interior, it was adorned by Charlemagne "with gold and silver, with lamps, and with lattices and doors of solid bronze. He had the marble columns for this structure brought from Rome and Ravenna."

Scholars attracted to Charlemagne's patronage of learning, which included generous stipends, journeyed from all over Europe. From central Italy came the religious poet Paulinus of Aquileia and the grammarian Peter of Pisa. From north Italy came the Lombard scholar Fardulf, originally taken as a hostage during Charlemagne's Lombard conquest; Fardulf later became a Charlemagne loyalist and was named abbot of St. Denis in northern France. Others came as exiles from Moslem-occupied Spain.

But the most important scholar of all who came to Aachen was Alcuin of York (732–804), trained at Jarrow by Bede's students. Praised by the Frankish chronicler Einhard as "the greatest scholar of the day," Alcuin wrote widely on religious subjects, arranged votive masses for days of the week, corrected the unrefined Latin of the Franks' religious texts, and standardized a new lower-case alphabet unknown in ancient Rome (and which you are reading right now). Alcuin served as Charlemagne's personal tutor between 781 and 796, as this largely untaught barbarian chieftain made an admirable attempt to educate himself in between battles and campaigns. "The king spent much time and labor with him studying rhetoric, dialectics," says his enthusiastic aide and chronicler Einhard, "and especially astronomy; he learned to reckon, and used to investigate the motions of the heavenly bodies most curiously, with an intelligent scrutiny."

This all sounds marvelous—except that it was not entirely true. Indeed, the emperor's reign fell far short of the grand renaissance he dreamed of, and which some historians have claimed. Medievalists today insist Charlemagne's intellectual accomplishments were mostly superficial, the pastime of a bright but unrefined warlord who treated learning as a precocious child might admire a shiny stone or delight in trying to work out a riddle or a puzzle. The emperor, these historians say, built libraries and filled them with manuscripts, but treated them as treasured ornaments, like fine cloth or rare spices—objects of status rather than texts to read and learn from. Of course, he was hardly alone in this attitude during an age when even supposedly learned monks spent lifetimes endlessly copying manuscripts that few understood or bothered to read closely. As for his clocks, Charlemagne considered them to be little more than toys, exquisite playthings that gave him a veneer of high culture when in reality his own artisans and scholars lacked the knowledge and skill to design and construct anything approaching the great water clock of Sultan Harun al-Rashid.

Charlemagne seems to have collected scholars in much the same way. As a *barbari* fascinated by these symbols of a sophisticated culture, he did not entirely comprehend them but hoped to emulate them nonetheless. Even worse, most of these scholars were barely educated themselves. In 809, two decades after Charlemagne issued his edicts ordering that children be educated, this was proven when a legal proceeding at Aachen summoned the greatest experts in the empire on ecclesiastic time reckoning. These "experts" were questioned in regard to Charlemagne's orders to teach *computus* throughout the empire, but it is obvious from the record of the proceeding that they had little understanding of this science. Dressed in the medieval academic's dark, heavy robes and felt hats, paid for by the emperor, these men of learning sadly did not grasp even the basics of Bede's mathematics and calculations—or much else.

Charlemagne himself, educated as a warrior in the centuries-old tradition of Germanic leaders and kings, could barely read and could not write despite years of lessons from Alcuin and Peter of Pisa—and despite the insistence of Einhard that the emperor had mastered astronomy and time reckoning. "He . . . used to keep tablets and blanks in bed under his pillow," admits Einhard, "that at leisure hours he might accustom his hand to form the letters; however, as he did not begin his efforts in due season, but late in life, they met with ill success." Most of his nobles were entirely illiterate. Nor could most of his scribes and scholars except Alcuin write in decent Latin.

Another exception was Einhard himself, who wrote a reasonably clear, notably secular history of Charlemagne's era. He also seems to have been more keenly aware of the intellectual shortcomings of the imperial court than other would-be scholars of his day. "I, who am a barbarian," he tells us, "and very little versed in the Roman language, seem to suppose myself capable of writing gracefully and respectably in Latin." He also complains that his history will be derided by both those who clung to the writings of the ancients and "despise everything modern" and those who despised all learning, including "the masterpieces of antiquity."

In such an environment it was all but impossible for true scholarship to flourish. Nor was it a place and a time where the calendar was likely to be fixed, even as it now drifted against the solar year by almost seven days since Caesar's reform.

In 800 Charlemagne accepted the title of Holy Roman Emperor from the pope, an event that signaled the Church's acknowledgment of what had been the political reality in Europe since at least the beginning of the Moslem conquests: that St. Peter's could no longer depend on either local Germanic kings in Italy or the Byzantines to protect Christendom in the West. Lacking armies and political power, the popes had long been leaning toward the Franks as their new protector. Charlemagne had cemented this relationship in 774 when he crushed the Lombards, who then ruled the northern half of Italy, bringing the still nominally independent papal territories under his protection. This added a measure of security for the prelates at St. Peter's, though politics in Rome remained tumultuous enough that sixteen years after driving away the Lombards Charlemagne again found himself leading troops to Rome to aid a pope besieged not by an army but by powerful local factions in the

chaotic city. In a potent demonstration of the Church's frailty as an earthly power, Pope Leo III was waylaid in 799 in Rome, where Einhard says his enemies "had inflicted many injuries . . . tearing out his eyes and cutting out his tongue."

Charlemagne's response was characteristically decisive. In November of 800 he marched on Rome, restoring order so swiftly that a grateful Leo proposed a novel reward that sharply underscored the dependency of the Church on the Frankish royal house: naming Charlemagne emperor of a new "Holy" Roman empire. This was an astute political move by the newly blind and dumb Leo, fusing the secular might of Charlemagne with the formidable religious power of the Church, an update of Constantine's fusion of the Roman imperium with the Church some five centuries earlier. Charlemagne reportedly resisted the crown at first, supposedly out of modesty, though unlike Caesar when Mark Antony offered him the diadem eight and a half centuries earlier, Charlemagne did not refuse the crown when it was publicly offered during a mass at St. Peter's on Christmas day, 800.

Neither Leo nor Charlemagne may have realized it at the time, but this crowning was not merely an act joining a desperately weak pope with a powerful patron. It also acknowledged and reinforced two enormous changes in Europe that would profoundly affect all aspects of life over the next several centuries, including the calendar and the science of time reckoning.

First was the consolidation and victory of the Catholics in finally eradicating virtually all other sects in the West, as all Christians fell in line behind their rules for everything from dating Easter and punishing heresy to when it was acceptable to have sex. The second was formalizing the rising new political and economic order in Europe we call feudalism. Though still unformed and incomplete when Leo placed the jewel-encrusted gold diadem on the long white hair of the Frankish king, the rough outline of the fiefs, duchies, baronies, and royal domains were then taking shape in a system that would dominate Europe for centuries—with the Church as an integral component, both as a huge feudal landowner and as a legitimizer of sovereigns who as a class would henceforth claim their rule was sanctioned by God.

In this way the princes of Europe and the pope essentially agreed to a pact that gave the Catholic Church authority over all religious matters—including most science—backed by the power of the princes and their gendarmes and armies. At the same time the Church provided the princes a potent religious undergirding to support their authority; and an all-pervasive code of conduct that would comfort their subjects with its message of hope and redemption, while keeping them under tight control.

Obviously this "pact" was another body blow to any scientific endeavor that might challenge a dogma purporting to know the truth in all matters, including time. It also meant that anyone who presumed to suggest reforming the Latin calendar would have to go to St. Peter's rather than to kings and princes—something no one dared attempt until Roger Bacon tried it four and a half centuries later.

If the emperor of the Holy Roman Empire was illiterate and treated time as a game and timepieces as toys, then what did time mean to a farmer in the Rhine River valley in the year 800? What sort of calendar did, say, a weaver use in central France? Or a fisherman on the often drizzly coast of Bede's Northumbria?

Little is known about commoners during a period when even chronicles and official records of kings and nobles are scarce. On a continent of illiterates barely getting by, most people seem to have spent their days hoeing fields, avoiding wild beasts, worrying about crops and the weather, burying the dead, celebrating marriages and local saint's days, and telling stories around hearth fires during the long, cold, deadly winters. They lived, ate their meager portions, bore children, repaired leaks in their thatch roofs, tried to avoid armies if they came into the area, took an excited peek at the lord or king if he came along their road, grudgingly paid taxes, attended mass, followed the orders of the lord's foreman, and died, all in a continuous cycle of days and years that to them had no discernible past or future.

Most Europeans lived in isolated rural communities, ignorant of the wider world. For instance, archaeology reveals that most people in Britain lived in farmsteads either by themselves or in little clusters. The latter were not even real towns; they were more like settlements of wood-plank or sod huts thatched in straw. Few towns, in fact, existed, or would exist until later in the Middle Ages, when groups of farmers banded together to form villages, and local squires and lords gathered their peasants into communal-style systems for agriculture. Some lords were beginning to organize their large estates into reasonably efficient units, some worked by slaves and others by serfs. But the transition from the chaos of the barbarian era to true feudalism was barely underway.

In 800, cathedrals, castles, and local administrative manors for kings and nobles were the most highly organized communities in western Europe. This is where craftsmen, tradesmen, servants, and beggars congregated, though in small numbers since there was little work—or spare change—for these classes. Even a "city" such as London—described by Bede as "an emporium of many peoples coming by land and sea"—was really just a larger than average cluster of fading Roman stone buildings, a small port, and a community that shipped a few slaves and possibly some wool in exchange for luxury items, metals, and a scattering of other products from the continent that few could afford.

To us, the world of the farmer on the Rhine and the weaver in France would have been one of dust and foul smells and mostly unhealthy-looking people wearing crude wool tunics, leggings, and loose-fitting leather shoes, or no shoes at all. During the day they worked from dawn to dusk in backbreaking manual labor when crops had to be planted, tended, and harvested; in the off season they had less to do. At night they slept in straw-topped huts in compounds shared with farm animals and heated with fires and stones baked hot during cold winter days.

In Charlemagne's time and throughout the Middle Ages, over half the children died before age five. Life expectancy was only 35 years. Farming methods were crude, with wooden hoes, sticks, and little knowledge of fertilizer or systems of crop rotations. This meant famines were frequent and often deadly. Even in good times the diet was poor: barley or millet with a few vegetables in gruel served daily with a piece of stale bread and an occasional slice of cheese or fruit. Epidemics raged across districts and kingdoms every few years. Between 540 and 600, six known plagues struck major Mediterranean cities in the East and West, wiping out many, many thousands of people. Most feared was smallpox, apparently first seen in Europe in 451 when Attila's warriors became stricken before a coalition of Romans, Ostrogoths and Franks defeated his Huns in France at the crucial Battle of Catalaunian Fields. Russian folklore also warns about kissing the Pest Maiden, and those who knew the Bible lived in fear of the fourth horseman in the Book of Revelations, sitting on his "pale horse . . . and his name that sat on him was Death."

Thieves and bandits ran amuck in 800, though there was little to steal outside the well-guarded estates, cathedrals, and small walled towns. Poems and stories from that long-ago era tell of a great fear of wild animals; dark, haunting forests where no one dared venture; and imaginary beasts and devils with fiery eyes and horns. People were earthy and pragmatic, but in the absence of scientific explanations for why the sun rose and fell and countless other mysteries they were also highly credulous and susceptible to even the most ludicrous superstitions and rumors. In 810 a buzz spread across Frankland that an enemy of Charlemagne was poisoning cattle with a magic dust. Another rumor insisted that "cloud-borne ships" manned by "aerial sailors" were on their way to ravage the land. Even the sensible Bede offhandedly describes dozens of miracles occurring within living memory of his own time—such as the curing of the blind man by Augustine, the archbishop of Canterbury, while meeting with the heretic Celts under the oak tree on the border of Kent and Wales.

Few people in this world had a need for formal calendars. Like Hesiod's Greeks and pastoral cultures around the globe, Europeans in the age of Charlemagne were primarily interested in predictable cycles and cues from nature. Chaucer, for instance, starts *The Canterbury Tales* with a calendric guide to the seasons and crops that Hesiod would have understood perfectly:

> *Whan that April with his shoures soote*
> *The droghte of March hath perced to the roote,*
> *And bathed every veyne in swich licour*
> *Of which vertu engenred in the flour;*
> *Whan zephirus eek with his sweete breeth*
> *Inspired hath in every holt and heeth*
> *The tendre croppes, and the yonge sonne*
> *Hath in the Ram his half cours yronne,*
> *And smale foweles maken melodye . . .*
> *Than longen folk to goon on pilgrimages.*

Anyone living in England during Chaucer's age would have instantly understood the references to April as the time of "shoures soote"—"showers sweet"—and of "zephirus eek with his sweete breath," referring to the west wind that blows sweetly after the "droghte" of March. Indeed, Chaucer considers it vitally important to establish first in his reader's mind the time of year when "folk to goon on pilgrimages," though he evidently has little interest in the actual year or date beyond noting that this is April: the start of spring.

And why should he? He was writing for an overwhelmingly agricultural people closely connected to the soil, for whom time was more than anything a powerful constant: a progression of youth and old age, birth and death, and as always the rise and fall of the sun each day. Nothing symbolized this better than the medieval wheel of fortune that perpetually turned, with one's lot sometimes up and sometimes down in a never-ending cycle. This great wheel of life represents the insecurity of an age when death and disaster lurked everywhere, and it explains to a large extent the mind-set of resignation about progress and change that deeply permeated this culture, caught in a constricting, repeating, seldom-altered circle of time.

Fused onto this world were the cycles and time schemes of the Church, the most obvious religious time marker being the weekly observance of Sunday, the day of rest and worship, which remains today the most constant time marker in a Christian's religious life. Next came the regular progression of saint's days. These saints came in two varieties: the major saints and apostles, whose days were marked by feasts and ceremonies, and the hundreds of lesser saints, whose days were marked by a reading of their lives in monasteries and perhaps a prayer by someone looking for a special favor connected to that saint's cult.

By 800 the number of saints with their own day of remembrance and worship numbered in the many hundreds in some areas. Because Rome would not formalize the process of achieving sainthood until centuries later, the Northumbrians in Britain were essentially free to declare their own saints; so were the Lombards, Burgundians, Bavarians, and Irish. They often became a matter of fierce local pride and identity, such as St. Patrick in Ireland and St. Andrew in Scotland. Some became potent national symbols as people began to think of themselves as Irish, French, Scottish, and Basque.

In Charlemagne's time saint's days were called "birthdays"—*genethlios*, or *natalis* in Latin—which then meant simply "commemoration," and which comes from the custom in pagan Rome of celebrating deified rulers on a particular day.★ Typically a saint's day fell on the date of

---

★The actual word "birthday" comes from an Old Norse word, *burdardagr*, and the German *geburtstag*.

his or her martyrdom. Complete lists of saints began appearing in the fourth century, describing in often gory detail their burnings, hackings, crucifixions, mutilations, and drownings; the particulars of place and time; and the locations of a saint's relics—a fragment of bone, a tooth, or a swatch of hair. Many of these holy men and women became identified with some important attribute. St. Nicholas became the patron saint of children and virgins; he was also revered by sailors before becoming the model for our St. Nick, or Santa Claus. His day is December 6. St. Agnes's day is on January 21; she is still venerated by many Catholic women "for her chastity and purity." And St. Giles, a seventh-century bishop whose lameness has made him revered as the patron saint for anyone who is ill or disabled, has his day on September 1.

People prayed to these and other saints for good crops, rain, and healthy children almost as pagans once prayed to specific gods assigned to oversee agriculture or fertility. For many these cults offered an intimate faith tied to a real person whose holy life had given them special powers in heaven to intervene either directly in a person's affairs or as a supplicant to God on that person's behalf. In this way the calendar of the saints became both a progression of religious dates and festivals and a highly personal cycle of time in which cultists eagerly anticipated certain days, which they marked with gifts and prayers.

The major saint's days became so well known and widely observed that many people used them in place of Caesar's scheme of months and days. A farmer would tell his friends that he last roofed his hut not on March 21, but on St. Benedict's day,* and he would remember that his second child was born on St. Augustine's day, not on August 28. Likewise, travelers in the Middle Ages talked about arriving in Rome or Paris on the Feast of the Assumption or during the feast of St. Stephen rather than on a numbered date. "He passed over the bar of Sanlúcar on Sunday, the morning of Saint Lazarus, with great festivity," writes a chronicler of a sixteenth-century conquistador leaving Spain on a ship bound for the Americas. The Christian calendar of

---

*St. Benedict's Day was changed to July 11 in 1969 to avoid Lent.

saint's days and festivals was also used to name places. Florida—"flowers" in Spanish—was named after the day Juan Ponce de León arrived there in 1513, on the day of the Pascua Florida, the "feast of flowers" to celebrate Easter.

To remember all of these saints, monks and priests wrote poemsongs listing each one and why he or she was revered, which they memorized and frequently repeated. One of the earliest of these is a verse calendar of saints from Britain, the "Metrical Calendar of York." Penned in the late eighth century, the same century Bede wrote his *History*, it contained the names of some 81 saints, many of them now so obscure that a line or two in this calendar poem is the only information about them that survives. Who, for example, is St. Cletus? Or St. Linus? One wonders whether over the centuries, as memories and details of these forgotten saints grew dim, the monks chanting their names even knew who they were:

> *At its beginning November shines with a multi-faceted jewel:*
> *It gleams with the praise of All Saints.*
> *Martin of Tours ascends the stars on the ides.*
> *Thecla finished her life on the fifteenth kalends.*
> *But Cecilia worthily died with glory on the tenth kalends.*

A better and certainly more entertaining method for preserving dates and details about the saints evolved into a new literary form in Middle Ages: the martyrology, books that dated saint's days and described details of their lives. Bede, for instance, wrote a classic martyrology with 114 entries, researched with his usual thoroughness. He also uses our modern system of assigning a number to each day of the month, rather than the kalends and ides used in the contemporaneous "Metrical Calendar of York," yet another indication of the many different dating schemes then in use. A typical entry in Bede:

23 November, at Rome, the feast of St. Clement, the bishop who, at the Emperor Trajan's request, was sent into exile in the Pontus.★ While there, because he converted many to the faith through his miracles and teaching, he was cast into the sea with an anchor tied to his neck. But as his disciples prayed the sea receded three miles, and they found his body in a stone coffin within a marble oratory, and the anchor lying nearby.

Writing and copying calendars of saints' lives became a major focus of scholars and artists during the Middle Ages. Every morning monks read descriptions of that day's saints. Even today a large department at the Vatican stays busy keeping track of the thousands of officially recognized saints and the thousands of others who have been canonized or beatified as steps toward possible sainthood.

Saint's days in the Middle Ages remained an informal method of dating for centuries, though scholars and kings preferred more formal systems for dating edicts and compiling chronicles. In the centuries following the collapse of Rome this tended to be the Roman scheme of kalends, nones, and ides, though as the empire became a more distant memory Europeans began replacing it with a number of alternatives. As we know, Bede and Charlemagne embraced our own system of *dies mensis,* where the days of the month are counted in a simple numeric order from 1 to 30 or 31. Others used a variety of other methods, including one called the Bologna custom, practiced widely in Italy, which counted days from the first to the middle of the month, but then started counting backward toward the last day of the month. Another scheme used verses in a poem in which each Latin syllable represented a day of the month. For example, in one of these poems the verse for the first 17 days of January ran: "*Cisio Janus Epi sibi vendicat Oc Feli Mar An,*" with *Ci* corresponding to January 1, *si* to January 2, and so forth. The idea was that people could memorize the verses—which usually commemorated appropriate local saints— and would then know the proper day in its proper order.

★The Black Sea.

But few farmers on the Rhine or weavers in France ever stopped to think about such things. To these people, who had little control over their environment or their lives, the whole idea of attempting to calculate and measure something as unfathomable and unremitting as time was either blasphemous or laughable. The few written insights into the mind-set of commoners on the subjects of time, calendars, and science in general suggest a great deal of snickering at monks, scholars, and astrologers bumbling about counting on their fingers and staring at the sky. The Miller in Chaucer's *Canterbury Tales* pokes fun at an astronomer-astrologer, but the verse might also have applied to anyone with their head in the clouds, so to speak:

> *Men sholde nat knowe of Goddes pryvetee*
> *Ye, blessed be alwey, a lewed man*
> *that noght but oonly his bileve kan!*
> *So ferde another clerk with astromye,*
> *He walked in the feelds, for to prye*
> *Upon the sterres, what ther sholde bifalle,*
> *Til he was in a marle-pit yfalle.*

In other words, this "clerk," or scholar of astronomy, did not heed what every "lewed"—unlearned—Christian knew and believed: that "men should not know of God's private affairs." He foolishly studied the moon and stars and was such a dolt, according to the Miller, that he was looking up when he should have been looking down at his feet—and fell in a "marle-pit."

Still, even the simplest Christian presumably had at least a vague knowledge of critical events in Christian history. In fact, for most people this timeline remained far more real than a history of their own era: the sequence of the Creation and events in the Old Testament;

and episodes in Christ's life and the lives of the saints. These events needed to be recorded and dated to become valid, and it was this need that motivated time reckoners such as Dionysius and Bede to devise their year-by-year dating schemes in an age when otherwise few people cared about what year it was beyond year 6 or 10 in the reign of their local king or squire.

Several chronological schemes were proposed and used besides Dionysius Exiguus's *anno Domini*. These included the old Roman system of fifteen-year interdictions, which had started with the first year of Constantine's reign in 312. Iberians used something called the Era of Spain, which tracked Easter cycles starting with the Roman conquest of Iberia in 38 B.C. Others observed the Era of the Passion, with year 1 dated back to A.D. 33, supposedly the date of Christ's crucifixion and resurrection. But none was as popular as a possible alternative to the year of our Lord than a timeline based on the date of the Creation as year 1. Bede, for instance, carefully studied what he considered the relevant passages in the Bible and somehow came up with a specific day that he believed God began forming the sky, earth, and water: March 18, 3952 B.C. If Europeans had decided to use Bede's calculation of the Creation, our year 2000 would be 5951 A.C.—after the Creation.

And what about predicting the future? Time in Christianity was, of course, heading someplace: to Christ's second coming and eventually to eternity, events that would occur along the same timeline as past events. This made it tempting for medieval chronologists to try to date not only the beginning of the world, but the end. A century before Charlemagne, one scholar in the royal Frankish court calculated, using poor addition, that the world was 5,928 years old in the year 727. Applying this to the notion that the world was moving through six ages of 1,000 years apiece, this computor decided that the world would end in exactly 72 years.

Bede, following the example of Augustine, condemned such predictions. He insisted that future time belonged to God, "who, as the Everlasting, created times whenever he wanted, knows the end of times, and puts an end to the fluctuating processes of time when he wishes." Still, most people who thought about such things believed that however old the earth might be, the end was near. "The world is growing old," wrote Fredegar, a seventh-century Frankish chronicler who wrote in corrupt Latin. "We live at the end of time."

Medieval chroniclers were constantly looking for portents of the grand finale: plagues, earthquakes, eclipses, battles, and omens of every kind. Mystics looked for signs of the Antichrist's coming, with writers such as the remarkable theologian and poet Hildegard von Bingens offering vivid descriptions of what he would look like: "A beast with monstrous head, black as coal, with flaming eyes, wearing asses' ears and with gaping jaws decorated with iron hooks."

Amidst this official pessimism certain dates took on meaning at least for a few, such as the coming of the year 1000, though the *anno Domini* system was still not widely followed.★ Even where it was, Christians did not necessarily *fear* the end. They expected trials and tribulations and a final, horrific apocalypse, as predicted in the Bible. But they also looked forward to what would come after the current age ended and the calendar truly stopped—when Christ would usher in an age of eternal happiness for the elect, which of course included them.

Meanwhile, as Christians waited for Armageddon they had more immediate concerns: they lived, ate, worked, bore children, sang songs, laughed, cried, and died as they always had, with only an occasional thought about the Antichrist or the last days of a calendar most medieval Europeans were at best vaguely aware of.

★Actually the first millennium came in the year 1001, since there is no year zero in our calendar.

But despite the "vast indifference to time" that permeated Europe during the reign of Charlemagne, already under way were real changes that centuries later would usher in a revolution in the perception of time. For even though Charlemagne saw clocks as curiosities, his keen interest in them and the idea of telling time made a lasting impression on future generations. At the same time a new invention was spreading slowly across the West: the bell. Called *glocka* in German—whence came our word *clock*—bells were used to signal hours and other times of the day. By legend, church bells were invented in the fifth century in the town of Nola in Campania—thus the term "Campanola bells." Another legend credits Pope Sabinianus (pope from 604 to 606) with ordering churches to mark the hours of the day by ringing their bells. Bells probably spread first to monasteries, where monks used hand bells to signal canonical hours. Later tower bells summoned people to mass.

Bells probably had a minimal impact on the average person. Yet they were the first mechanical "clocks" to govern everyday life in Europe, usually rung according to time as measured on a water clock or sundial. Imagine a farmer in a field being told to have an acre plowed by the time the bell tower rang noon, when before he had been told by his lord simply to work until the sun was high. Or think of a clock that signaled the beginning of a mass with an exactitude never before known when hours were measured using the position of the sun in the sky. This was an entirely different way of viewing time, with a measurement of it being assigned a specific value.

The Holy Roman Emperor Charlemagne died on January 28, 814. His empire died soon after, as his heirs argued and fought and divided his realms among them. With it perished the political order that Charlemagne had briefly imposed. So did the emperor's infatuation with learning, manuscripts, and marvelous timepieces, which it turned out was not shared by his immediate successors. They dismissed the scholars from the court and closed the schools for children opened by the emperor. Still, the age of Charlemagne ignited a spark, with the scholar Alcuin of York and others compiling encyclopedias and collecting manuscripts. It also provided an example and a context for quality, taste, humanistic culture, and sound grammar, which laid a foundation for a slow—very slow—evolution toward an era when dates and calendars would begin to matter to more than just a few monks sitting in their cloisters trying to calculate the age of the world and when the end would come.

But Europe was not where the action was for time reckoning and the calendar anyway in Bede's or Charlemagne's era. Indeed, as Europe slept, developments were under way far to the East, where science was not ignored and a long line of brilliant thinkers were making discoveries that centuries later would penetrate at last the darkness of the West to astonish and inspire men like Roger Bacon, and to once again commence the movement of time.

8

The Strange Journey of

365.242199

There are also others who know something.
—SEVERUS SEBOKT, SYRIAN BISHOP, A.D. 662

In 476, far away in time and place from Charlemagne's dark, imposing castle at Aachen, beyond the eastern border of Frankland and on across the Balkans, the territories of Byzantium, and the vastness of Mesopotamia and Persia, a Hindu genius was born on the Ganges River. A blend of Ptolemy the astronomer, Pythagoras the mathematician, and Bacon the rebel, Aryabhata was one of a remarkable group of Indian scholars, and a pivotal figure on one of the stranger journeys ever taken by an assemblage of ideas across time and geography.

This saga of ideas begins six thousand years ago in Mesopotamia and Egypt. It then moves to ancient Greece, only to hopscotch to India during the great Hellenistic surge that accompanied Alexander's armies in the fourth century B.C. The ideas then arc back west centuries later, landing in the great centers of Islamic learning after the Arab conquest of Persia and India. The Arabs in turn carried the knowledge to portals in Spain, Syria, and Sicily, where it made its way into Europe, to be embraced at last by pre-Renaissance thinkers such as Bacon.

During the journey each culture that seized on these new ideas

137

added significant contributions, and together over the centuries they assembled a remarkable body of learning about mathematics, astronomy, and other fields of science and art that would eventually make it possible for time reckoners in Europe to correct Caesar's calendar—and to measure time with an accuracy essential to propel science into the modern world.

Aryabhata himself was a key figure in a tradition in India stretching back to at least 1500 B.C. when light-skinned Aryans—the ancestors of those who later founded the Hindu religion—swept down from the northwest to conquer an earlier civilization, the Harappa.

The Aryan-Hindus started writing about mathematics as early as 800 B.C., when their priests began laying out complex designs for temples and altars, and for dividing up land—a process that led to the discovery of the basic geometric rules that also seem to have marked the first stages of advanced cultures in Egypt, Sumeria, China, and Central and South America. Hindus called their version of this property- and construction-inspired math *sulvasutras*—*sulva* being the name of cords used by architects to mark off a structure's foundations, and *sutra* referring to rules governing a ritual or science.

These crude concepts were written down in Sanskrit verse, and were critical to an early understanding of shapes and their relationship to one another—including versions of the Pythagorean theorem and early geometric algebra.★ Eventually they turned this body of knowledge skyward to measure the planets and the stars, which led to sophisticated

---

★The Pythagorean theorem is one of the most fundamental concepts in mathematics. It is critical for making basic astronomical observations for anyone wanting to use the stars or sun to measure time. The theorem says that in any right-angled triangle, the square of the hypotenuse is equal to the sum of the squares of the other two sides. It is named for the Greek Pythagoras (sixth century B.C.), though several cultures discovered it independently.

attempts to measure time, including astrological predictions of the future based on the movements of the sun and of the zodiac.

The age of the *sulvasutras* ended around A.D. 200, during a period of political instability that lasted until the early fourth century, when the Gupta dynasty seized most of northern India and launched Hindu India's classic age. Taking up where the *sulvasutras* left off, Gupta astronomers in the fourth and early fifth centuries made great strides in mathematics and astronomy, recording them in a series of texts known as *siddhantas,* or "systems" of astronomy. Written in the two hundred years before Aryabhata began working, they provided him with the universe of fundamental concepts he used for his own work—including estimates of pi, basic rules of trigonometry, the motion of the planets and stars, and the length of the year.

Aryabhata grew up during the final years of the Gupta golden age, when India was a world center of art, science, literature, and architecture. Learning was considered a sacred duty, and educated Hindus were expected to know not only the basics of reading, writing, and numbers but also to be adept at poetry, painting, and music. This was the age of the *Kama Sutra,* the text that treats love as a fine art, offering alongside lovemaking positions a list of "arts to be studied, together with the *Kama Sutra.*" These include swordplay, composing poetry, "playing on musical glasses filled with water," chemistry, teaching parrots to speak, grammar, tattooing—and mathematics.

Gupta India was hardly a paradise for everyone. Governed by a strictly enforced caste system, the poor endured a life of crushing poverty similar to that in many Indian villages today, little changed since Aryabhata's day—crowded clusters of straw-thatched huts, dusty markets filled with burlap sacks of rice and peppers, and lean men trudging to and fro to work small plots of land. Still, excavations in Gupta centers attest to the large numbers of merchants, artisans, and others in a large middle class who enjoyed a prosperity on par with the golden age of Rome, which had been a major trading partner with Gupta India until its collapse. Archaeologists sifting through Gupta ruins have found heaps of coins and blown glass from Rome; and from Roman sites as far away from the subcontinent as Pompeii, others have unearthed Indian statuettes, vases, mirrors, and busts of Roman men with Indian hairstyles.

Aryabhata's birthplace is unknown. Nor does anyone know what he looked like, though he himself tells us he lived in the busy imperial capital of Kusumapura. Today the city is a hot and hauntingly quiet stretch of drooping palms, buzzing flies, and crumbled ruins that extend some 12 miles along the banks of the Ganges, near modern Patna. This is in northeastern India, 250 miles north of Calcutta and just 100 miles south of the sudden rise of the Himalayas. At its height the city was filled with throngs of people: beggars disfigured by disease, rich traders in white robes, musicians playing cymbals and flutes, silk-clad Brahmans averting their gaze to avoid making eye contact with someone from a lower caste, and priests with hair tinted by henna toting statues of gods and goddesses. Enormous, airy palaces lined the Ganges, alongside imposing conical temples studded with statuary and ornaments. The entire city was shrouded with a gauzelike veil of incense, smoke, and dust.

A leading instructor at a school near Kusumapura, Aryabhata spent most of his life collecting and compiling everything ever written in India about the stars, geometry, numbers, and time reckoning in his magnum opus, the *Aryabhatiya,* a slim volume written in Sanskrit verse. And while only 123 metrical stanzas long, it packs an enormous amount of information in what became a handy volume of mathematical and astronomical concepts passed down and commented on over the centuries. Some of it is highly accurate, some not, a contradiction that prompted a famed Arab mathematician named Ibn Ahmad al-Biruni (973–1048) to comment that Hindu mathematics offers two types of nuggets: common pebbles and costly crystals.

Aryabhata starts his poem with an invocation to Brahma, "who is one in causality, as creator of the universe." He then divides his work into three parts: on mathematics *(ganita),* time reckoning *(kalakriya),* and the sphere *(gola).* In the section on time reckoning Aryabhata describes the Hindu calendar, including measurements of the months, weeks, and year, and various time spans relating to Vedic mythology over the course of millions of years. In the section on astronomy he estimates the length of the solar year at 365.3586805 days, some 2 hours 47 minutes and 44 seconds off from the true year in Aryabhata's era, which equaled 365.244583 days.* He also gets the diameter of the earth almost right, at 8,316 miles, but is wildly off on his estimated orbits of the sun, moon, and planets. Aryabhata believed that the earth was a sphere that rotated on its axis, and he understood lunar eclipses as the shadow of the earth falling on the moon. Some historians have even detected what one calls "glimmerings in his system . . . of a possible underlying theory" that the earth might revolve around the sun, a possible nod toward the truth about our heliocentric solar system a thousand years before Copernicus.

In his section on math Aryabhata gives formulas for the areas of a triangle that are correct, and areas for a sphere and a pyramid that are not. He calculates pi to be 3.1416, another near hit that is so close to the value given by Claudius Ptolemy some three hundred years earlier that it is possible Aryabhata was influenced by the great Alexandrian astronomer, though no direct link is known. Aryabhata wrote a famous stanza giving his value for pi originally in Sanskrit verse:

> Add 4 to 100, multiply by 8, and add 62,000. The result is approximately the circumference of a circle of which the diameter is 20,000.

Unfortunately Aryabhata does not explain how he arrived at his formulas and calculations. Nor does he offer proof for what ends up being a catalogue of arbitrary rules. One senses that the *Aryabhatiya* was intended as more of a supplement or summary for people already familiar with the concepts than a comprehensive encyclopedia or the-

---

*Because the tropical year is steadily slowing, the year in Aryabhata's era was slightly shorter than our current year of 365.242199 days. The difference between then and now is about seven seconds.

ory of mathematics. He may have written down details elsewhere in a work now lost, or perhaps as exercises for his students.

Aryabhata is also credited with writing a work called the *Khandakhadyaka,* which means "food prepared with candy," possibly referring to the pleasure it gives. But the original has been lost. Only a heavily edited and annotated version exists, reworked by another renowned Indian mathematician, Brahmagupta (598–665).

A debate has long simmered over where Aryabhata's ideas and the corpus contained in the *sulvasutras* and *siddhantas* came from. Indian historians have long insisted that it sprang up purely as the product of indigenous genius, with origins possibly going back to the dawn of civilization on the Indus River in about 2500 B.C. This is when the ancient Harappa culture began to flourish in cities made of mud brick that since have all but crumbled away, making them difficult to learn about. Still, archaeologists have unearthed evidence—building designs and measuring devices—that suggest the enigmatic Harappa did master fundamental mathematical principles. Possibly these were passed on to the Aryan-Hindus who stormed down from the north to conquer the Harappa and seize most of northern India, though the history of this period is so murky that no concrete link can be made.

A more definite influence came from Greece after 326 B.C., when Alexander seized northwest India. In his wake came the concepts of Pythagoras, Meton, Eudoxus, and Alexander's instructor, Aristotle. The conqueror's armies also stirred up and brought with them the scientific knowledge of other cultures swept into his brief empire, including Egypt and Mesopotamia. The Greek hegemony in northwest India lasted only a few years, falling apart soon after Alexander's death in 323 B.C. But Greek knowledge and culture lingered as Greek traders established thriving enclaves in India and established the lucrative trade routes west that persisted throughout the Hellenistic and Roman eras.

This allowed Indians an opportunity to absorb Greek ideas about planetary theory and geometry. One of the *siddhantas*, the Paulisha Siddhanta, may even be named after a minor astrologer from Alexandria, Paulos Alexandros (fourth century A.D.). Certainly this work contains striking similarities to Claudius Ptolemy's trigonometry and astronomy, on which Paulos based his work—including a value for pi nearly identical to that later identified by Aryabhata.

The Chinese may have been another influence. They maintained a vigorous enough commerce with India that the two cultures swapped styles of clothing and architecture and even words. This was particularly true after Buddhism spread across the Middle Kingdom in the late Han Dynasty and during the period from 220 to 589, known as the Six Dynasties. No direct evidence exists of mathematical ideas being transferred between China and India, though the lifetime of the great Chinese mathematician Tsu Ch'ung Chi (430–501)— who came up with the most accurate estimation of pi in the world until the European Renaissance—overlaps with that of Aryabhata, who wrote his *Aryabhatiya* two years before Tsu's death. Tsu also measured the precise time of the solstices, building on the work of another brilliant Chinese astronomer and court astrologer, Zhang Heng (A.D. 78–139), who corrected the Chinese lunar calendar in the year A.D. 123 to bring it into line with the seasons. Tsu also proposed reforms in 463 to China's lunar calendar, which apparently were rejected.

But no influence in India was apparently more significant for our calendar—and the mathematics and equations needed to fix it—than that of another cradle of civilization: the Tigris-Euphrates Valley in Mesopotamia. Or so it seems, despite the lack of evidence for direct links between India and Mesopotamia on matters of mathematics and the calendar. For instance, no manuscripts exist to tell the tale of an Indian scholar visiting ancient Sumer in such-and-such year. Still, many of Vedic India's mathematical and astronomic concepts seem strikingly similar to some of those used in the Near East, such as the *sulvasutra* rules for construction using Pythagorean triads—which appears in Babylon before it does in India. Other shared concepts include ideas about fractions, algebra, polygonal areas, and applied geometry that appear first in Mesopotamia and later in the *sulvasutras* and *siddhantas*.

It seems inconceivable that ancient mathematicians and time reckoners from Ur and Harappa, and later from Babylon and Vedic India, remained completely ignorant of one another during the many centuries of commerce between the Tigris-Euphrates region and India. Some Indians almost certainly picked up a little cuneiform, the writing of Mesopotamia for four thousand years, perhaps from watching a Babylonian merchant scribbling down figures on a wax tablet on the coast of Sind, or from a Mesopotamian ship captain calculating wages to pay his porters in Gujarat.

However the contact may have occurred, it seems likely that somewhere over the millennia the Mesopotamians sparked an idea that led to one of the great mathematical discoveries in history: the system of arranging numbers that mathematicians call "positional notation," now used by virtually the entire world. Among many other things, this made an accurate calendar and higher mathematics possible.

In positional notation, numbers are arranged in a sequence whereby each number stands for itself multiplied by a base number that increases by one power of the base with each place. For instance, in our base-10 system the number 365, representing a rounded-off approximation of the year in days, is drawn from a set of ten symbols, 1-2-3-4-5-6-7-8-9-0, that are arranged to increase tenfold with each place. So we have 3 hundreds (10^2) 6 tens (10^1), and 5 digits (10^0), the digits referring to the original source of the base-10 system—counting with one's fingers.

It is a concept so central to our modern system of numbers—and our way of life—that we hardly think about it, though this was not the case throughout most of human history. Indeed, the only culture to invent a true positional notation system in preclassic ancient times was Mesopotamia, whose mathematicians stumbled on it almost four thousand years ago—predating all other cultures by millennia.

To fully appreciate the significance of positional notation and a number such as 365, one has to realize that for most of human history people used either their fingers or bulky, hard-to-manipulate symbols representing ever-increasing numbers.

The first written numbers seem to have been sticklike signs scratched onto bones or rocks long before written languages were invented. We still use a version of them today to count small numbers of things that accumulate over short periods of time: yellow-breasted warblers sighted during a morning hike; runs batted in during an afternoon baseball game; or the number of patients seen at a well-baby clinic every hour. For instance:

$$8 = \text{Ш} \; \text{III}$$

But this system quickly becomes impossibly cumbersome even with a number as simple as 365, roughly the number of days in a year:

Ш Ш Ш Ш Ш Ш Ш Ш Ш Ш Ш Ш Ш Ш Ш Ш Ш Ш Ш
Ш Ш Ш Ш Ш Ш Ш Ш Ш Ш Ш Ш Ш Ш Ш Ш Ш Ш Ш
Ш Ш Ш Ш Ш Ш Ш Ш Ш Ш Ш Ш Ш Ш Ш Ш Ш Ш Ш
Ш Ш Ш Ш Ш Ш Ш Ш Ш Ш Ш Ш Ш

It takes several minutes just to write out this number—never mind using it to add or subtract, or to write out and perform a more sophisticated calculation such as determining the angle of the earth to the sun, or the shape of a temple along the Euphrates or the Ganges. This led early civilizations to devise more compact system of symbols, often closely related to early forms of written languages. For example, the Egyptians invented a hieroglyphic-inspired sequence of numbers:

I	∩	ϑ	⌇	⟩	⌐	☉
1	10	100	1,000	10,000	100,000	1,000,000

And in the Americas the Maya used a system of lines and dots originally represented by sticks and pebbles, later adding hieroglyphic symbols to represent larger numbers:

•	—	◉
1	5	20

Other cultures such as those of the Greeks, Romans, and Chinese used letters of their alphabets to represent numbers:

Greeks:*

A	B	Γ	Δ	E	F	Z	H	Θ	I	K	Λ	M	N	Ξ	O	Π
1	2	3	4	5	6	7	8	9	10	20	30	40	50	60	70	80

ϟ	P	Σ	T	Y	Φ	X	Ψ	Ω	λ
90	100	200	300	400	500	600	700	800	900

Romans:

I	V	X	L	C	D	M
1	5	10	50	100	500	1,000

---

*When small letters were introduced into the Greek alphabet, these small letters replaced the older capitals.

Chinese:*

一 二 三 ◻ 五 六 七 八 九 十 ⼆ ⼆ ⺫ ⺫
1 2 3 4 5 6 7 8 9 10 20 30 40 50

⊥ ⊥ ⊥ ⊥ 百 千 萬
60 70 80 90 100 1,000 10,000

Here is how these cultures would have written the number 365,
the number of days in a solar year, rounded off:

in Egyptian   ⁊ ⁊ ⁊ ∩∩∩|||
in Chinese   三
               百
               六
               十
               五
in Maya   𝋡𝋡𝋡
in Greek   ΤΞΕ
in Latin   CCCLXV

These number symbols were a great improvement over sticklike
signs, but still presented problems for calculating or recording complicated equations and large numbers. This is why a positional system
was such a phenomenal breakthrough—a leap of inspiration made by
a long-forgotten Mesopotamian who undoubtedly became frustrated
with writing out large numbers. Perhaps he was a scribe assigned the
unenviable task of counting barrels of wine coming and going from
Ur's royal palace. Or an architect designing a ziggurat but running
out of space on a clay tablet as he made his calculations, and so he

---

*The Chinese developed a second system of numbers called "rod numbers" that use
a base ten position notation system using 18 number symbols instead of nine. Much
later, they added a zero—the first use coming in A.D. 1247.

|	||	|||	||||	|||||	Т	ПГ	ПГ	ПГ	—	=	≡	≣	≣	⊥	⊥	⊥	⊥
1	2	3	4	5	6	7	8	9	10	20	30	40	50	60	70	80	90

365 = ||| ⊥ |||||

invented a quick shorthand to save space. Here is what 365 looks like
in cuneiform with its positional notation:

𒌋𒌋𒌋𒌋𒌋   𒁹

In this system each 𒁹 stands not for a power of 10, but of 60, since
Mesopotamians used a sexagesimal system rather than a decimal one.
The smaller 𒁹 = 1; the Mesopotamians also used the symbol 𒌋 to
represent 10. This means that six 𒁹's equal (60×6), or 360, with five
small 𒁹's added to make 365—a cuneiform number that is certainly
easier to write out than 365 in, say, Egyptian. Still, even cuneiform
could be unwieldy. Indeed, ancient Sumerians and Babylonians often
had to contend with writing out long strings of repeating symbols for
each digit from 0 to 60, rather than our simple symbols of 1 through
9, and 0. When Babylonian astronomers calculated the length of a
lunar year versus a solar year, the equation rounded off would have
looked like this:

365 days	𒌋𒌋𒌋𒌋𒌋	𒁹
−354 days	𒌋𒌋𒌋𒌋	𒌋𒌋𒁹
11 days	𒌋	𒁹

It was this problem of unwieldiness—and more—that the Indians
solved by inventing our system of nine numeric symbols sequenced
in positional notation, later adding zero for the tenth symbol.

How exactly the Vedics of India figured out this brilliantly simple
scheme is another mystery, though they might have been inspired to
transform Mesopotamia's base-60 positional system into their own
base 10. Some historians also speculate that a connection exists between Indian numbers and ancient Chinese rod numerals, which also
have symbols for 1 through 10 used—after the third century A.D.—
in a positional system.

Whatever its origin, these symbols that eventually became our own

first appear in carvings on stone columns across north India as early as
250 B.C. or before, when Hindu mathematics was making the transition to a positional system. Written out in the early Hindu script
known as Brahmi, the first nine numbers look like this:

$$- \quad = \quad \equiv \quad \curlyvee \quad \sqcap \quad 6 \quad 7 \quad \varsigma \quad \rangle$$
$$1 \quad 2 \quad 3 \quad 4 \quad 5 \quad 6 \quad 7 \quad 8 \quad 9$$

Subsequent versions in the evolution between the earliest Brahmi
numerals, circa 250 B.C., and the numbers used by Aryabhata seven
centuries look like this:

Modern Number	Progression of the Centuries	Version in Use, c. A.D. 500
1	‒ ‒ ⌐ ‵ ? ?	?
2	= ↗ ㅗ ㅗ ㅗ	?
3	☰ ☲ ⩘ ⅔ ?	?
4	+ ⋎ ↳ ⅄ ⅄ ⅄	?
5	ⱶ ⱶ ⱶ	?
6	℮ Ɛ	?
7	? ? ?	?
8	? ⅄ ⅄ ↲ ⊂	⊂
9	? ? ? ? ? ⅌⅀	?

But this system still was not purely positional. Brahmi, lacking a
zero, also had individual symbols or groups of symbols to represent
10 through 90, 100, 500, and 1,000. In Brahmi the number 365 is:

$$7 \equiv \dashv \Gamma$$

The evolution from this version of Brahmi to a ten-digit positional
notation is not entirely clear. Historians suspect that the motivation

to drop the Brahmi symbols beyond the number nine came from the demands of the Hindu religion, which uses a calendar encompassing enormous spans of time to date its creation myths. These form a religious chronology stretching back millions of years, requiring the manipulation of large numbers—which is far easier when one uses powers of ten. The Indian use of counting boards also encouraged the development of number symbols that were simple and few in number.

The timing is also uncertain. Aryabhata was certainly aware of positional notation and apparently used it in his day-to-day calculations. But because he wrote his treatises in metrical verse he used words and letters to represent numbers—the equivalent of us writing out "twenty-nine" rather than 29—in an attempt at mathematics as poetry.

The first known use in India of the nine-digit positional system has been discovered on a plate and dated to the year 595. The number is a date—346—written in a decimal positional notation.

The first outside mention of the Hindus' system of nine numbers comes in 662 from the Syrian Severus Sebokht, an academic and bishop who lived in a Greek community founded a century earlier by scholars fleeing Athens after Justinian closed Plato's Academy, which he had accused of fostering paganism. Apparently Sebokht became piqued at his colleagues' disdain for any knowledge beyond the Greek sphere. Writing about the Hindus, he cites their "subtle discoveries of astronomy . . . their valuable methods of calculation, and their computing that surpasses description. I wish only to say that this computation is done by means of nine signs."

But nine does not make ten, meaning the system was not complete without zero, a concept critical to understanding the advanced mathematics needed to create an accurate calendar. Zero developed as Indians using the nine numbers for calculations found themselves needing to keep an empty column on their counting boards to represent "nothing," an idea they transferred to writing out numbers by leaving a space. But this could be confusing, since a space could mean either an empty position in a single number or the space between two separate numbers. To avoid confusion somebody along the way decided to make something of "nothing."

Who was first to scratch out a symbol for zero is yet another mystery. In Mesopotamia a symbol for the empty position appears late for this ancient civilization, arriving around the time of Alexander's invasion or just after, represented by two small wedges placed obliquely:

$$\lambda = 0$$

At roughly the same time or shortly thereafter the Indians began using a dot, a symbol that became widespread enough by the sixth century that the Indian poet Subandhu used it as a metaphor in his poem Vâsavadattâ:

> And at the time of the rising of the moon with its blackness of night, bowing
> low, as it were, with folded hands under the guise of closing the blue lotuses,
> immediately the stars shone forth . . . like zero dots . . . scattered in the sky
> as if on the ink blue skin rug of the Creator who reckoneth the total with
> a bit of moon for chalk.

The Indians referred to this "nothing"-dot as *sunya*, meaning void or empty. Our word *zero* comes from *sifr*, the Arabic version of *sunya*, which medieval Europeans altered to *ziphirum* in Latin.

Greeks of the classic age had no symbol for zero, because their numerical system did not require a zero place. But they were aware of the concept of a number that stood for nothing. Indeed, Aristotle rejected it as a nonnumber to be ignored, since one cannot divide by zero, or divide zero by itself. Nevertheless, Eurocentric scholars long assumed that the symbol for zero was invented by the Greeks, with no proof at all, speculating that it came from the Greek letter omicron—O—the first letter in the Greek word *ouden,* meaning "empty." But this unwarranted belief that Indians could not have come up with such a basic concept has given way to recognition that ancient Greeks did not really use such a symbol for zero, and that Indian mathematicians seem independently to have invented the dot and then the round goose-egg symbol. The first use of this symbol for zero in India appears in the year 876 in an inscription found in the Gwalior region south of Delhi, containing two numbers with zeros:

50: ౿౦   270: ౨౭౦

This comes two centuries after Severus Sebokht's mention of the *nine* Hindu numbers, though archaeologists have found the round symbol for zero in two numbers in an inscription in Malaysia—the numbers 60 and 606 as ౬౦ and ౬౦౬—that dates to A.D. 684. The Malay peninsula was then under Indian influence. Some historians also believe a treatise on mathematics known as the Bakhshali Manuscript may have been written as early as the third century A.D. It contains numbers with zeros and a fully developed decimal place-value system. The numbers include:

330: ౩౩౦   846,720: ౮౪౬౭౨౦

The first use of zero as a fully formed number seems to have appeared around the time of Brahmagupta in the seventh century, when this great Indian mathematician tried, but failed, to explain how zero could be divided by itself. The Maya also invented a true zero in about the third century A.D., using several symbols, including a half-open eye—◉—which they used to indicate missing positions as they wrote out numbers to represent time intervals in their calendar.

This explanation of zero does not quite finish our story about the mathematics needed to correct the calendar, since the year is not 365 days long, but 365.242199 days, give or take a few seconds. In other words, we have this pesky fraction to contend with, expressed here as a *decimal* fraction. This concept—and the ease with which we are able to represent this value—also did not come easily or all at once. Beyond the simplest divisions of a whole number, fractions posed a huge problem for humanity through most of history.

How do you divide three sacks of grain among five people? And how do you split up a year, month, day, hour, or minute into smaller parts?

Predictably, the first written symbols for fractions represented only the simplest divisions. The Mesopotamians used ⬥ for what we write as ½. The Egyptians used ⌒ for ½, and X for ¼, the X probably showing how an object was cut into four quarters. For other simple fractions using one as the numerator, Egyptians wrote symbols that look something like our modern system:

$$\frac{1}{5} = \text{⬚} \quad \frac{1}{12} = \text{⬚} \quad \frac{1}{20} = \text{⬚}$$

Romans organized their fractions around a division of one into 12 parts. This grew out of their system of weights, which was based on a unit of measurement called an *as,* split into twelve *uncias.* Symbols were assigned to each fraction, with I equaling an *as* (one whole), an S equaling ½ (six *uncias*), a = − for ¼ and a − equaling ¹⁄₁₂ (one *uncia*). To write out smaller fractions, Romans divided the *uncia* into 24 *scrupuli* and each *scrupulus* into 8 *calci,* and so forth. Each of these smaller groupings had their own symbols and names, such as ⊦⊣ for ¹⁄₉₆, called a *drachma,* and ℒ for ¹⁄₂₃₀₄, called a *calcus.* Several modern words are derived from this system—*ounce* and *inch* come from *uncia*, and *calculus* may come from *calcus*.

But these symbols are far too cumbersome and imprecise for sophisticated values and calculations. For instance, it was relatively simple for a Roman—or Bede, or Alcuin in Charlemagne's court—to write out the Latin whole number and fraction for the length of Caesar's year, which is CCCLXV $=$ $-$ days—365¼. But try writing the true solar year of 365.242199 days—the equivalent of 365 242,199/1,000,000—in Roman numerals. No symbol exists in Latin for such a precise number, a reality that profoundly affected the pursuit of determining an accurate year. Nor is it possible to calculate in Roman numerals a value that takes into account variations in the earth's motions, including the gradual slowing of the tropical year over the centuries.

As long as time reckoners used the Latin system—or Greek, Egyptian, or any other numerical system that lacked precise fractions—they were forced to conclude that it was impossible to calculate a true year. This powerfully reinforced the belief in the Middle Ages that if such a number existed, it was known only to God, when in truth the number was simply beyond the capability of the symbols and numerical system in use at the time—and continued to be until the thirteenth and fourteenth centuries, when Europeans began broadly adopting the earliest versions of the modern decimal system.

The idea of using decimal fractions came to Europe from the Arabs, though they were not the first to use positional notation to write out and determine fractions. Again this distinction seems to belong to the Mesopotamians, who over the millennia figured out a fraction system based on their own positional notation scheme—which gave them a precision and computing power far beyond that of any other system until the European Renaissance. But because Mesopotamia's system was based on 60 and not on a more manageable number such as 10, their remarkable discovery was limited by the complexity of carving into clay and stone place-values in negative powers of 60, which not only are indivisible for some fractions but also quickly become long and complicated symbols to write out. For instance, the length of the year in cuneiform numerals is:

$$365.242199 = \text{𒐏𒐐 𒁹 𒌋𒁹𒐏𒁹𒐊𒁹}$$

Which breaks out in base 60 as:

$$365 \quad + \quad .233333 \quad + \quad .008611 \quad + \quad .000255$$
$$6(60) + 5 + \quad 14(60)^{-1} + \quad 31(60)^{-2} + \quad 55(60)^{-3}$$

The Chinese by the third century A.D. had also discovered how to write fractions using their positional notation, and did so using our familiar base-10 system. But their discovery does not seem to have traveled beyond the Far East. As for the Indians, for some reason they did not develop decimal fractions, despite having base-10 positional notation for whole numbers. Instead they devised an early version of placing one number over another to represent fractions—a numerator over a denominator—that was apparently borrowed from Greek mathematicians in Alexandria, with one difference: they placed the denominator over the numerator. The bar line was introduced later by Arab mathematicians.

Of course, the vast majority of people in ancient times had little use for fractions beyond the most simple divisions of a whole. Only a handful of mathematicians and astronomers cared to be more precise—and even they tended to simply round off numbers either to the closest simple fraction or to the nearest whole. This is undoubtedly why early astronomers, from Hipparchus and Ptolemy to Aryabhata, were able to note that the 365¼-day year was wrong, but seemed willing to accept this rounded-off number as tolerable enough that none called for a correction, or for reforms in the official calendar.

When Aryabhata wrote his *Aryabhatiya* in 499, at the precocious age
of 23, Gupta culture and learning remained at a high point. But even
as he pondered pi and the position of planets, a dark cloud was fast
engulfing the empire: the Huns. This eastern branch of the scourge
that had hastened the crash of Rome had for years been hammering
away mercilessly against the Gupta frontier to the northwest.

By the time the *Aryabhatiya* appeared, the Huns had broken
through the Guptas' main defensive lines to devastate parts of north-
western India. But unlike Rome, the Guptas, with help from the
Chinese in the north, had weakened the military strength of the
hordes over the years to the point that the invaders were unable to
thoroughly conquer the Indians or destroy their culture. During the
middle third of Aryabhata's life the Huns set up a shaky kingdom that
ruled from modern Afghanistan to central India, never reaching Kusumapura. Aryabhata lived long enough to see a coalition of Indian
kings and warlords drive them back into Kashmir in 542, when he
was 66 years old. He also had lived long enough to see the golden age
of Gupta culture slowly eroded, even if the continuity of Indian culture was preserved.

As the political situation worsened, the spirit of open inquiry and
free thinking that had thrived earlier was squelched by a turn to conservative Vedic values. This apparently got Aryabhata into some trouble with his more controversial theories, particularly his supposed hint
that the earth might circle the sun. At least this seems to be the case
given the vigor with which later Indian scholars, perhaps anxious to
conform to the more rigid orthodoxy of the day, dismiss this theory
less on academic than religious grounds.

How Aryabhata responded to his critics is unknown. But we have
a clue to his true feelings, and his willingness to express them, in a
short passage at the end of the *Aryabhatiya*. It reads like something

Roger Bacon would have written as a fevered defense of science. "He who disparages this universally true science of astronomy," says Aryabhata, "which . . . is now described by me in this *Aryabhatiya,* loses his good deeds and his long life."

But unlike Bacon, Aryabhata was revered by scholars and laymen alike, during and after his lifetime. Every great Indian mathematician and astronomer who came after him used the *Aryabhatiya* as the basis for their work and acknowledged his contributions. This includes Varahamihira (505–587), a contemporary of the elderly Aryabhata★ who wrote an encyclopedia that cites the master of Kusumapura, but emphasizes astrology over astronomy—a choice Aryabhata would have rejected as unscientific.

The great mathematician Brahmagupta (598–665) also held Aryabhata in high esteem, incorporating some of the earlier master's works into his own—and unfortunately editing them and adding his comments to the point that it is hard to tell what belongs to Aryabhata and what to Brahmagupta, since the originals Brahmagupta worked from have been lost. Brahmagupta's reverence did not extend to Aryabhata's controversial ideas. Nor did it stop him from offering corrections in his *Brahmasphuta-siddhanta,* written around 628, to what he considered his predecessor's mistakes on matters ranging from the altitude of the sun's ecliptic to Aryabhata's measurement of the diameter of the earth.

Aryabhata's impact was so profound in his homeland that in 1975 modern India honored this ancient genius by launching a scientific satellite named the *Aryabhata* on an Indian Intercosmos rocket. Unlike the ideas of its namesake, the satellite failed after only four days and came crashing back into the atmosphere on February 11, 1992.

After Brahmagupta, India continued to produce noted mathematicians, including Bhaskara (1114–1185), considered by mathemati-

★There may have been two Aryabhatas working at roughly the same time, Aryabhata the Elder and Aryabhata the Younger.

cians to be the most brilliant in his field anywhere during the twelfth century. But he was the last true standout in medieval India.*

All of these men contributed mightily to the evolution of concepts that three centuries after Aryabhata's death would continue their journey to the West via a people that in Aryabhata's era were primitives barely known to the great civilizations of the day. Living on a vast desert to the south of the empires of Persia and Byzantium, they began stirring to life only in the final years of Brahmagupta's life, then suddenly they burst out of their desert peninsula to begin the conquest of much of the Near East and southern and central Asia. In the process they discovered and then embraced the ancient knowledge of India, Greece, and Mesopotamia, creating an unlikely amassing of ideas drawn together in what became the early medieval era's greatest center of learning: Baghdad.

---

*In 1887 another mathematics genius was born in India, Srinivasa Ramanujan, who tragically died at the age of 33. His natural fluency and intuition with numbers has been compared to the free-ranging and eclectic style of thinking of Aryabhata and other earlier Hindu mathematicians.

9

From the House of Wisdom to Darkest Europe

It was He that gave the sun his brightness and the moon her light,
ordaining her phases that you may learn to compute the seasons and the
years. God created them only to manifest the Truth. He makes plain his
revelations to men of knowledge.

—THE KORAN, C. A.D. 630

In 773, some 250 years after Aryabhata's death, a delegation of diplomats from the lower Indus River Valley arrived in the new Arab capital of Baghdad. Dressed in brightly colored silks, turbans, and glittering gems, this group probably traveled by sea from the Indus delta around the desert coast of modern-day Iran and up the turquoise waters of the Persian Gulf to the port city of Abadan—some 30 miles inland now because of silt built up over the centuries. They would then have sailed up the Tigris about 200 miles to Baghdad, passing by the hot, dry banks lined with tiers of ancient, irrigated terraces and stone cities dating back to the time of Sumer and Ur, arriving at last outside the gates of al-Mansur's magnificent city.

A half century after the Arabs had conquered the lower Indus River Valley, in 711, this delegation was one of many dispatched by local Indian authorities to the court of Caliph al-Mansur to provide him with news about their province, and to settle outstanding disputes. They also hoped to impress the great caliph, the founder of the Abbasid dynasty, with the richness and sophistication of their country by showering him with gifts—perhaps a gem-encrusted suit of armor, a flute carved out of ivory, a highly prized falcon, or a silk tapestry depicting scenes from their province.

This particular delegation also brought with them an astronomer, undoubtedly having heard that al-Mansur was not only a mighty general and military ruler, but also a patron of the arts and sciences. The astronomer's name was Kanaka. An expert on eclipses, he reportedly carried with him a small library of Indian astronomical texts to give to the caliph, including the *Surya Siddhanta* and the works of Brahmagupta (containing material on Aryabhata). Nothing more is known about this Kanaka. The first known reference to him was written some five hundred years later by an Arab historian named al-Qifti.

According to al-Qifti, the caliph was amazed by the knowledge in the Indian texts. He immediately ordered them translated into Arabic and their essence compiled into a textbook that became known as the *Great Sindhind* (*Sindhind* is the Arabic form of the Sanskrit word *siddhanta*).

No one is sure if this incident per se ever happened. But something like it must have, in order to bring the works of India into the sphere of the early Islamic scholars, whence they would travel to Christian Europe through Syria, Sicily, and Arab-controlled Spain. A version of the *Great Sindhind* would be translated into Latin in 1126. This was one of dozens of critical documents that would contribute to the knowledge base needed to propel Europe into the modern age, and to calculate a true and accurate year.

Kanaka allegedly visited the court of the caliph in Baghdad about a century and a half after one of the most extraordinary moments in history: the sudden maelstrom that came out of Arabia in the mid-600s. Driven by a potent fusion of religious zeal and a centuries-old martial tradition among the tribes of the desert, the armies of the Prophet Mohammed were at first a phenomenon of arms and religion, though they soon became an unlikely force for the advancement of learning. This came in part from the Prophet's command that the faithful seek knowledge, but also because the Arabs did not follow the example of the *barbari* in the West, who had looted and destroyed the cities and provinces of Rome. Instead the Arabs assimilated the cultures of the peoples they conquered—much the same way the early, uncouth Romans had done centuries before when they eagerly embraced and absorbed the cultures they conquered in Greece and the Near East.

In a sense the Arabs arrived just in time. Most of the ancient centers of learning, and the cultures that had nourished them, were in a state of exhaustion or outright collapse by the mid-600s, after decades of warfare and internal decay. To the East the Gupta era was ending as India broke up into small kingdoms and struggled to fend off fresh onslaughts from the Huns; in the Near East a long war fought between Byzantium and Persia ended with a peace treaty in 628, leaving both empires gravely weakened; to the west the *barbari* continued to battle over what was left of Rome.

Not surprisingly, this period produced little original thinking and was a low point in intellectual output from the Himalayas to the British Isles—with some notable exceptions, such as Brahmagupta in India and a few scattered scholars still struggling to work in the Greek tradition within the Byzantine Empire. But even there the output was meager as the rump of the old Roman Empire, pressed by enemies on all sides, had become more stridently orthodox. Indeed, for decades the imperium and Church had been repressing rival Christian sects, pagans, and anyone else who did not fall in line behind an increasingly strict religious dogma—including scholars.

This religious retrenchment in Byzantium had begun under Justinian in Cassiodorus's time. In 529 he had closed the nine-hundred-year-old Academy of Plato in Athens and had dispersed its scholars, claiming it was a hotbed of paganism.★ Fearing for their lives as well as their intellectual freedom, many of these scholars had fled to Persia,

---

★Most scholars date the end of the ancient Greek culture to the closing of the academy in 529.

where they established a kind of Academy in exile. This was a pale imitation of the original, though this community of scholars remained viable enough that when the Arabs seized Persia a century later these Greeks were able to play a major role in bringing the texts and learning of the ancient Hellenes to the attention of Arab scholars.

The events leading up to the meeting between al-Mansur and Kanaka began modestly. In 610, roughly 30 years after Cassiodorus's death in faraway Italy, a 40-year-old merchant in the desert oasis and trading post of Mecca claimed to have seen the archangel Gabriel in a vision. Commanded by the angel to lead a movement to purify and complete the religious tradition of Judaism and Christianity, Mohammed began to preach a simple message to the pagans in his town: one of total submission (which is what the word *Islam* means in Arabic) to one god: Allah.

At first only his family and a few friends responded favorably. Nearly everyone else laughed at him, eventually forcing him and a tiny band of followers to flee Mecca in 622 for another desert oasis, nearby Medina. This later became known as the "Year of the Migration" (*hijra* in Arabic, *hegira* in English), which is the starting point of the Moslem calendar—a calendar Mohammed later insisted should remain purely lunar, to differentiate it from the lunisolar calendar of the Jews and the solar calendar of the Christians.

The Medinans welcomed Mohammed as a sagacious leader and arbiter of disputes, and he shrewdly used this reputation to build a power base. This allowed him to eventually unite the entire peninsula of Arabia under his authority and to organize a potent new military force inspired by his new religion of self-sacrifice and devotion to God. By 630 he had conquered his former hometown of Mecca, where the people now embraced his religion.

Then Mohammed died on June 8, 632.

His death threw his followers into a state of confusion, but only briefly, as one of Mohammed's most important disciples, his brother-in-law Abu Bakr, took over as the first *khalifat rasul-Allah*—"successor to the apostle of God"—or *caliph*. This did not settle the leadership crisis then or afterward. But it did allow the Arabs to take advantage of their newfound unity and their religiously inspired warriors to launch a rampage of conquest that within two decades of the Prophet's death crushed the armies of Persia, overran Egypt, Syria, and parts of Asia Minor, and nearly took Byzantium.

In a second wave of conquests from 696 to the 720s the armies of Islam pushed north to the Caspian Sea and Turkestan, northeast into modern Iran up to the Aral Sea, and even briefly into Kashgar, on the edge of China's sphere of influence. To the southeast they conquered the lower Indus Valley. In the west they seized North Africa and raged into Spain, turning back only when they reached France and were confronted by a powerful Frankish army led by Charlemagne's grandfather, Charles Martel.

By the mid-eighth century the expansionist military force of Islam was largely spent for the moment, and the Arabs began to take stock of what they had conquered—politically, economically, and culturally. Having come from a desert where few were literate and the lifestyle modest, they brought little material culture to the ancient civilizations now under their sway. Their significant contributions were language and religion, and this is where their talent as master assimilators came into the fore, as they seized on the clothing, dress, architecture, philosophy, literature—and science—of the Persians, Greeks, and Indians they now ruled.

The possibilities offered by this crucible of cultural interaction burst forth just over a century after Mohammed's death when al-Mansur built his magnificent new city as a symbol of his awesome power and of learning. Urbane and sophisticated, al-Mansur and the early Abbasids lavished the wealth and power of their empire on science and the arts. The Arabs' golden age of literature, architecture, and science, centered in Baghdad, reached its apex during the reigns of al-Mansur's successors Haroun ar-Rashid (ruled 786–809) and his son al-Mamun (ruled 809–833). This was when the Indian texts first brought by Kanaka, and the others that came later, were translated, organized, and studied along with the knowledge of the ancients from Greece and Persia, and eventually synthesized into the forms that would later reach Europe.

Sir Richard Burton, the nineteenth-century explorer and orientalist, compared Baghdad during its glory years to the Paris of his century, a city that would also have rivaled Rome at its height. But the truth is that no one will ever know for sure the splendor that was Baghdad, for it was utterly destroyed, almost to the last brick, first during a period of civil war among the later Abbasids, and then in 1258 by an invading Mongol army.

At the core of Baghdad was a massive inner city of palaces, administrative buildings, and army barracks. Known as the Round City, it stretched for two miles in diameter, and was ringed by three concentric walls. In the center stood the caliph's Golden Palace, on an axis where four highways radiated outward in cardinal directions to the four corners of the empire. Surrounding the Round City were suburbs that grew up in all directions. These included sections for Jews and Christians, considered adherents of sister religions to Islam, and several large compounds for monasteries built by the Nestorian sect of Christianity. Banished two centuries earlier by Justinian, the Nestorians brought to Baghdad stacks of Greek scientific texts, which they helped translate into Arabic.

Someone walking through the hot, sunny streets of this Mesopotamian city in 800, the same year that Charlemagne was crowned emperor in the half-ruined city of Rome, would have passed a profusion of people—beggars, slaves, artists, thieves, merchants, and government officials dressed in a mix of styles from Persia, Greece, and

India; soldiers swaggering in polished armor; and traders from as far away as Spain and China.

According to an Arab chronicler named Abu al-Wafa Ibn Aqil, writing in Baghdad during the mid-eleventh century, the city was filled with palaces, gardens, fountains, and mosques of exquisite beauty—as well as hospitals, schools, and libraries. Along the Tigris, says Ibn Aqil, the rich built elegant residences, riding to and fro in small boats "in good trim . . . with beautiful finery and marvelous woodwork." He spends pages describing the great *suqs,* or markets, and their busy streets set aside for shoemakers, flower vendors, tailors, moneychangers, swordsmiths, perfumeries, and other merchants selling everything imaginable. One of these *suqs,* he says, was "unequaled for the beauty of its architecture," with "tall buildings with beams of teakwood supporting overhanging rooms." Another was known as "the meeting-place of learned men and poets." Still others offered shops full of books and entertainments ranging from recitations of the Koran to fencing and wrestling shows.

Scholars, engineers, scientists, and artists flocked to Baghdad from every corner of the empire, and were honored and well paid. Many came bearing manuscripts, and the early years of the Abbasid period became a great era of translation. This project was made infinitely simpler when the first paper factory opened in Baghdad in 794, using a process the Arabs learned from a Chinese prisoner captured during the 712 conquest of Samarkand, in modern Afghanistan. This invention would be passed on to Europe centuries later, just in time to provide late medieval scholars with an easy-to-make and inexpensive material on which to write out their own translations of ancient works.

As the translations and the originals began to stack up in the universities and libraries of Baghdad, al-Mamun ordered a museum and library complex built that became known as the House of Wisdom, the

Bait al-hikma. Completed by 833, it became the most outstanding single repository of knowledge and scholarship since the great library in Alexandria: a place where scholars pondered the ancient writings and, as time went by, developed theorems, concepts, and applications of their own.

By the second decade of the ninth century, just a generation after Kanaka's arrival, a new and vibrant Arab intelligentsia were making breakthroughs in everything from medicine, chemistry, and optics to a new philosophy of science that framed the pursuit of knowledge in terms of better serving God.

In the realm of time reckoning and astronomy, the Arabs first applied Greek and Indian ideas about measuring time to a practical need for their religion: when exactly they should kneel and pray, which Mohammed required all Moslems to do five times a day. This inspired early Arab astronomers to use and improve upon Greek instruments such as the astrolabe, sundial, and globe to better calculate the angles of the sun at various times of day. Astronomers also advised architects throughout the Moslem world where to build mosques so that the faithful could follow another command of the Prophet—that they always face the direction of Mecca when they pray, whether they are in the Hindu Kush or on the Rock of Gibraltar.

Moslem astronomers and mathematicians also went to work refining the Islamic calendar. This calendar—whose year 1 began in our year A.D. 622, when Mohammed fled Mecca for Medina—was established by the second caliph, Umar, around A.D. 634. Years in the Islamic calendar are indicated with the abbreviation A.H., which stands for the Latin *anno hegirae,* or "the Year of the Migration." Since then it has been running at the standard lunar time of 354 days a year, drifting across the seasons to start on the same day every 32½ years.

Each month in the Islamic calendar begins about two days after the new moon, when the first sliver of the crescent moon is sighted. Because the lunar month averages about 29½ days, Umar arranged the twelve months of the Moslem year to alternate between 29 and 30 days:

Name	Length in Days
Muharram	30
Safar	29
Rabi'u'l-Avval	30
Rabi'u'th-Thani	29
Jamadiyu'l-Avval	30
Jamadiyu'th-Thani	29
Rajab	30
Sha'ban	29
Ramadan	30
Shavval	29
Dhi'l-Qa'dih	30
Dhi'l-Hijjih	29

Most of these month names predate Islam, with some referring to seasons—suggesting that the Arabs' calendar may have been lunisolar before Mohammed's day. The second month, Safar, meaning "yellow," originally came around in autumn when leaves were turning color. Mohammed also designated four months that were sacred, when Moslems were forbidden to go to war or to conduct raids; of these, the ninth month, Ramadan, is the holiest, when Moslems are supposed to fast and abstain from sex during the daylight hours in order to learn self-discipline, and to concentrate on spiritual matters. Some believe the word *Ramadan* comes from the Arabic *ramz,* "to burn," because the fast is supposed to "burn" away one's sins. In the Koran Mohammed writes:

> As to the month of Ramadan in which the Koran was set down to be man's guidance
> . . . as soon as any one of you observeth the moon, let him set about the fast.

From this relatively simple early calendar Arab astronomers in the House of Wisdom and elsewhere worked to make the most precise lunar calendar possible. Their solution was a 30-year cycle of 360 lunar months, which is accurate against the true orbit of the moon to within a day of drift every 2,500 years. But this system requires frequent intercalations, with one day being added to the final month, Dhi'l-Hijjih, in the 2nd, 5th, 7th, 10th, 13th, 16th, 18th, 21st, 24th, 26th, and 29th years of each 30-year cycle.

To facilitate this and other practical astronomical inquiries, Caliph al-Mamun ordered an observatory built in Baghdad in 829, and one soon after outside of Damascus. Astronomers also set up a network of observation points across the empire that allowed them to conduct experiments. One of these set out to determine the size and circumference of the world, which Arabs assumed was round. Taking measurements on a plain north of the Euphrates and near Palmyra, astronomers were able to calculate the width of a degree of the meridian,★ coming up with 56⅔ Arabic miles. This is just 2,877 feet wider than the actual degree.

One of the astronomers involved in the project of measuring the distance between two meridians was almost certainly Abu Jafar Mohammed ibn Musa al-Khwarizmi (780–850), perhaps the greatest of the scholars working at the House of Wisdom during the golden age, and the most influential mathematician on any continent during the early Middle Ages.† As famed among Arabs as Euclid and Ptolemy, and later respected by the Europeans of Roger Bacon's day, al-Khwarizmi was probably born near the Aral Sea in modern Turkestan, called Khwarizmi in his era. Working in the city of Merv, south of the Aral Sea, he became famous enough to be summoned to Baghdad

---

★This refers to a measurement made by locating a meridian and moving north or south along it until one has moved exactly one degree of latitude.
†His name means "Mohammed, the father of Jafar and the son of Musa, from Khwarizmi."

in 820 by al-Mamun, who appointed him "first astronomer" and later head of the library at the House of Wisdom. An Arab version of what Europeans call a "Renaissance man," al-Khwarizmi wrote on a diz-zying number of subjects from mathematics and astronomy to geog-raphy and a history of the Arab caliphates. He also led three scientific missions to India and Byzantium to meet with scholars and collect manuscripts.

He is best known, however, for being one of the first major scholars in the Arab world to use the accumulating store of knowledge from India, Greece, and Persia to make his own discoveries. These include the invention of modern algebra. Indeed, the word *algebra* itself comes from one of al-Khwarizmi's books, *Kitab al-jabr wa al-muqābalah (Cal-culation by Restoration and Reduction)*. Later this became a standard text-book of mathematics in European universities until the sixteenth century. The word *algorithm*—*algoritmus* in Latin—comes from me-dieval Europeans' use of al-Khwarizmi's own name to refer to the study of mathematics.

Al-Khwarizmi wrote out the oldest surviving *zij*—set of astronomic tables—in the Arab world, much of it based on Indian charts possibly brought to Baghdad by Kanaka. This *zij* later made the journey to Spanish Córdoba and onward to the rest of Europe, where a Latin translation made in 1126 became one of the most influential works on astronomy in medieval Europe.

Perhaps most important of all was a small booklet al-Khwarizmi penned in 825. Called *Algoritmi de numero Indorum* when it was later translated into Latin, this short treatise detailed something the great sage of Baghdad apparently picked up from reading Brahmagupta: the numerical system of the Indians—the nine symbols and a placeholder called *sunya*. Amazed by the usefulness of these simple symbols and of positional notation, he demonstrated in his pamphlet their superiority to the Greek numbers then used in Baghdad, and to the cruder Bed-ouin numbers the Arabs had brought with them from the desert. At the time he wrote his booklet the "new" Indian symbols looked something like this:*

ı ٢ ٣ ٨ ٢ ٤ ٧ ٢ ٢

Later Arab mathematicians expanded on the system described in al-Khwarizmi's pamphlet and on the Indians themselves to take the Hindu idea of *sunya*—the Arabs' *cifra,* our *zero*—and use it not merely as a placeholder, but as a number like any other in certain calculations and equations. They also made a mathematical leap that the Indians did not, applying the system of positional notation to create decimal fractions—the first of which appear in an obscure book by an otherwise unknown Syrian mathematician named Abul Hassan al-Uqlidisi in 952 or 953. These discoveries made it possible just before the end of the first millennium of the Christian era to actually write out the number that represents the true solar year—365.242199 days—though as of yet no one had been able to come up with such an exact astronomic value. It also would have been written without the dot for the decimal point, which was added much later.

Al-Khwarizmi's contemporaries in Baghdad were delighted with his little book. Used to a dizzying rush of new knowledge in this era of learning and scholarship, they quickly dropped the old methods of counting and embraced the new—which greatly accelerated the development of mathematical theory that would lay the foundation for modern science, including the reform of the calendar.

Scholars in Damascus to the west began using the new numbers just a few years later, but this invention took almost a century and a half to make the long journey to Spain, Sicily, and other, more distant outposts of Islam. It took longer still to make the leap across the borders to a conservative Europe largely uninterested in new ideas,

---

*No one knows what the symbols looked like in al-Khwarizmi's booklet because no original has survived in Arabic. The only extant copies are Latin translations.

particularly those connected to a people they considered heathens in league with the devil.

Al-Khwarizmi was hardly the only genius at work in the Arab world during its glory years between the founding of Baghdad as the Abbasid capital in 763 and the final dissolution and fragmentation of the Islamic empire in the 1200s and 1300s. It is impossible to mention all of these, though a handful stand out above the rest in terms of the calendar. These include another denizen of the House of Wisdom born around the time of al-Khwarizmi's death, Abu Allah Mohammed Ibn Jabir al-Battani (c. 850–929), known in Europe as Albategnius. In a book called *On the Motion of the Stars* he expounded on Indian trigonometric methods to show that the distance from the earth to the sun varies during the year, something we now know happens in part because the earth's orbit is elliptic. Al-Battani also refined values for the length of the year by comparing it with calculations made by Ptolemy in 139. He came up with a figure that was 2½ minutes too short—but that was because Ptolemy had placed his equinox a day late. Had Ptolemy been correct, al-Battani's year would have been only half a minute short.

A half century later another Arab astronomer, Abu ar-Rayhan Mohammed ibn Ahmad al-Biruni (973–1048), was born in central Asia. There he thrived despite the growing instability in the region as the Abbasid caliphate collapsed and its territories fragmented into shifting emirates ruled by local shahs and warlords.

Before the age of 30, in the midst of wars between rival kings, al-Biruni was able to make extensive observations of equinoxes and to travel to and fro taking highly accurate measurements of latitude. Also before turning 30—even as he was forced at times to go into hiding because of politics—he managed to write at least eight works. These included a treatise on timekeeping, a timeline of past events dated according to the Moslem calendar, and arguments for and against the earth's rotating on its axis, taking up the debate of Aryabhata versus later Indian astronomers.

Al-Biruni later became a diplomat for one rival shah and was imprisoned by another, though he was eventually allowed to continue his work as he followed an invading Moslem army into India. There he learned Sanskrit and studied every ancient text he could find, compiling his findings into a book called *India*. This offers a remarkably candid and critical analysis of Hindu mathematics and the *siddhantas*. In his late sixties al-Biruni wrote a study on the specific gravity of gems; at the age of 80 he wrote an alphabetical guide to 720 drugs, listing each according to their names in five languages.

The year al-Biruni died yet another Arab scholar and poet was born, Umar ibn Ibrahim al-Khayyami (ca. 1048–1131)—known in the west as Omar Khayyám. Admired today outside of the Arab world exclusively as one of the greatest Islamic poets, Omar Khayyám was much more. Prolific in a number of fields, in mathematics he greatly expanded on al-Khwarizmi's algebraic principals and on Euclid's geometry; as an astronomer he spent 18 years working in an observatory in Isfahan, 200 miles south of modern Teheran in Iran, where among other things he measured the solar year at 365.24219858156 days. This was both accurate and overly precise considering the gradual slowing of the earth's rotation. Omar Khayyám also devised a solar calendar with eight leap years of 366 days every 33 years—a slightly unwieldy system that nonetheless was more accurate than the Gregorian calendar. Apparently he proposed his new calendar as a reform to his local shah in 1079. How this ruler responded is unknown.

In one of his famous versus from *The Rubáiyát*, Omar Khayyám offers a poet's assessment of what it means for a scientist to try to measure time—and the arrogance of those who blithely count and add and take away days on a calendar:

> *Ah, but my calculations, people say,*
> *Have squared the year to human Compass, Eh?*

If so, by striking from the calendar
Unborn tomorrow and dead Yesterday.

Another scholar working in the Islamic world, the Jewish astron-omer Abraham bar Hiyya ha-Nasi (1070–1136), wrote in Barcelona the first Hebrew work devoted exclusively to the study of the cal-endar, including a prediction based on the Torah of when the Messiah might appear. Yet another astronomer appearing very late in the classic Arab era was Ulugh Beg (1394–1449), the ill-fated son of a shah, who briefly ruled Samarkand and was put to death by his own son during a coup. Ulugh Beg gave a measurement for the length of the year that came to 365 days, 5 hours, 49 minutes, and 15 seconds—just 25 sec-onds too long.

Still, the Arabs came very close to calculating a true value for a year they did not use in their own religious calendar, a measurement few Europeans at the time even cared about. Even those who did care struggled with crude and incomplete formulas and data, a situation that seemed all but hopeless—and might have remained so if not for the eruption of learning out of Baghdad and other Islamic centers, a wave so powerful that it reached beyond even the distant rim of what was then the civilized world.

10

Latinorum Penuria
(*The Poverty of the Latins*)

*Why, as Bede himself admits . . . does a full moon appear in the sky
in most cases one day, and in others two days, before the computed date?*
—HERMANN THE LAME, 1042

No one in Baghdad during the heady days of al-Khwarizmi could have guessed that their work would help spark a revival of learning in Europe. A traveler trekking from the caliph's court to Aachen in the year 800 would have laughed at the idea that these foul-smelling *barbari,* ruled by an emperor who could not write, whose scholars copied old manuscripts rather than reading them, and whose mathematicians still counted with their fingers, would four centuries later produce a Roger Bacon. And three centuries after that a Copernicus.

Such a traveler would marvel at a people who had forgotten the mathematics, science, and philosophy first conceived by ancients from whom they traced their own cultural roots. He also would have smiled at the irony, if he had been able to predict the future, of a people who would one day rediscover the ancient knowledge they had lost in part from Arabic translations of the original European texts.

At first the process of transferring the concepts crucial to Europe's reawakening was almost imperceptibly slow. In 800 our adventurous Arab would have found at best several hundred ancient texts at Charlemagne's court, and a castle full of half-educated Franks still in awe of Bede. A scholar from Baghdad—or Damascus—arriving in

Latin Europe a century later, in 900, would have seen little difference. Even another century after that, in 1000, he would have witnessed only a few stirrings. Not until 1100 would our original traveler's great-great-great-great-great-great-great-great-grandson see any significant change, three centuries after Charlemagne tried—and failed—to rejuvenate learning in Europe.

Visitors checking the status of Latin time reckoning would have discovered roughly the same progression—computists in their monasteries during the 800s still worrying over lists of saint's days, updating Easter tables, and spending lifetimes trying to devise arcane systems for better measuring time. In Frankland an intrepid Arab traveler might have met the teacher, theologian, and scholar Rabanus Maurus (c. 780–856), a student of Alcuin and a prolific writer who spent many years of his long life fussing over how to divide the hour into ever smaller units: a useful idea, except that one has to ask why anyone in the ninth century would need to use, say, his *atom,* which he declared to be $\frac{1}{22,560}$ of an hour. Also, how would one measure the passage of such an infinitesimal moment of time with a water clock?

Other time reckoners during this period are now as notable for their unusual names as for their painstaking labors on *computus* and the calendar. They include three time reckoners whose work spanned the mid-ninth to the mid-eleventh centuries, all named Notker, and all who lived in the same Swiss monastery at St. Gall near Zurich. These were Notker the Stammerer, Notker the Peppercorn, and Notker the Thick-Lipped.

The 900s were little better than the 800s, with one major exception: a monk named Abbo of Fleury (945–1004), who advocated the use of water clocks that were more accurate than the sundial used by monks since before Bede. This allowed him to make slightly more accurate measurements than the venerable one for days, months, and years. Abbo also proposed a change in Dionysius Exiguus's chronology using *anno Domini,* substituting the old Latin style of passing from year 1 to year-1 to a timeline that added a placeholder in the zero position. To designate this "new" year he used the symbol for null, since zero itself had not yet reached Europe. This suggested change was ignored, however. So was his idea that the date for Christ's death as calculated by Dionysius was incorrect by some 20 years. But Abbo was an anomaly in a field that was becoming wearisome by the year 1000 with its rehashing of the same old formulas and arguments.

The last important work done in the traditional mold of *computus* and time reckoning was authored by yet another scholar-monk with an unflattering name, Hermann the Lame (1013–1054) of Reichenau, in western Germany near the border with Switzerland. Insisting early in his life that all scientific conclusions should be supported by "the insuperable truth of nature"—an astonishing admission in his day— Hermann used the recently arrived astrolabe and a special column sundial that he invented to compare what he saw in the sky to the fixed numbers used for centuries by computists. The first time reckoner since Bede to trust observation, he verified that the Church's calendar—including Easter and many feasts and saint's days—was out of sync with the cosmos. "Whence comes the error that the real age of the moon so often does not correspond to our reckoning, computus, or the rules of the ancients . . . ?" he asked in 1042.

Hermann's frustration was compounded by his repeated attempts to correct Bede and other computists, all of which failed to match up with what he saw in the sky. This left the lame monk of Reichenau wondering toward the end of his short life—he died at age 41—if the centuries-old tradition of *computus* and time reckoning was hopelessly flawed, based on erroneous assumptions about the movements of the sun, moon, and stars. But neither Hermann nor anyone else was willing to take this a step further and challenge the Church in an era when questioning St. Peter's was the same as doubting the Lord.

Hermann was hardly alone with his discomfiture. He was among the forerunners of a new breed about to come of age in a Europe finally shaking off its slumber. Men who would be raised and educated not in monasteries but in Europe's slowly reviving cities, where news of other cultures was arriving along with the first scatterings of long-lost texts by Greeks and newer writings by Arab and Indian scholars. Read and pondered, these would challenge not only the validity of old assumptions about the sun, moon, and the nature of time, but also the nature of the entire universe, including the role of man, and of God himself.

This new thinking would emerge during the eleventh and twelfth centuries in part because of a legacy set in motion centuries earlier by Charlemagne: the economic order he imposed. Far more lasting then his attempted renaissance of learning, feudalism by the twelfth century had long been the dominant system in west and central Europe, introducing a degree of stability unknown in the chaotic centuries after the collapse of Rome.

In 843, almost three decades after Charlemagne's death, the Treaty of Verdun had established the principle that "every man should have a lord." In theory, this meant that even the pope and emperor were subject to a higher authority—God—who sat at the top of what was later called the Great Chain of Being. Under this arrangement prelates came after the pope and monarchs after the emperor. Then came in the designated order bishops, priests, and greater and lesser nobles; and under them came squires, merchants, craftsmen, farmers, laborers on down to the lowest slave, and even to leafy plants, worms, and houseflies.

This hardly meant that politics in Europe around the year 1100 were serene. Kings, nobles, knights, squires, and occasionally bishops and popes fought among themselves almost as a birthright. Borders and dynasties shifted continually. Yet over the centuries since Charlemagne the basic outline of modern Europe had slowly evolved, with the states of France, Germany, and northern Italy emerging in the 900s, after a series of dynastic wars among Charlemagne's heirs.

To the south Christian princes had begun the long reconquest of northern Spain, capturing nearly a third of the peninsula from the disunited Moors by 1100. In the east missionaries had Christianized the Slavs, some of whom now called themselves Poles, Hungarians, Croats, Serbs, and Russians. The Vikings were giving up Thor and Wodin and settling down as Christian Danes, Swedes, and Norwegians, ending a two-century reign of terror against Britain and the coasts of northern Europe. In Britain William the Conqueror, duke of Normandy, seized England in 1066 and unified its realms, while the clans and tribes to the north joined to form the kingdom of the Scots.

The other big victor in the years since Charlemagne was the Catholic Church. By 1100 it reigned supreme, finally winning out over virtually all rival sects to achieve the monopoly of faith first envisioned by Constantine at Nicaea eight centuries earlier. In southern France and elsewhere religious malcontents made rumblings about the Church's all-too-secular emphasis on wealth and politics—and the propensity of some clergy to favor silks and gold over matters of the spirit. Scholars following the lead of Hermann the Lame and others also whispered in quiet corners of cloisters and cathedral schools that certain Catholic tenets concerning science and philosophy might be mistaken. But by and large the Roman Church was enjoying what would become under Pope Innocent III (pope from 1198 to 1216) the high-water mark of its power and influence.

The Church could horribly abuse its power, and seems to us centuries later to have been hopelessly dogmatic and repressive. But for the average Christian in 1100 Catholicism was mostly a huge comfort: a universal set of laws and beliefs that provided a powerful sense of spiritual unity and a deeply desired salvation, particularly for serfs and peasants—which meant just about everyone.

Indeed, the world was as arduous then as it had been for centuries. There had been a few improvements: the relative stability brought by feudalism; improvements in agricultural techniques, such as the invention of the heavy plow to use with horses; and increases in production that meant more food. But most people continued to live lives out of time toiling in fields and vineyards, repairing grass-thatched huts before the first winter storms, singing their children to sleep, suffering from rotten teeth, dying of measles and simple colds—an existence in which calendar time still did not matter and the seasons came and went in a never-ending cycle that few expected to change.

The major exception were the aristocrats, the great landowners who since Charlemagne's day had sat atop the feudal pyramid. Unlike everyone else, by 1100 they had seen their lives transformed, for the simple reason that they were fabulously rich. This privileged class had filled their coffers with gold and grain for three full centuries, growing even wealthier as production increased and the population in their fiefs and principalities expanded, with more wilderness cleared for growing millet, oats, cucumbers, grapes, figs, sheep, and cattle.

Aristocrats spent their newfound fortunes on thick-walled castles, private armies, gaudy suits of armor, falconry, flashy tournaments, feasts, and luxury goods imported from the East—silk capes, taffeta tunics, spices, and gems. Eventually this unbridled consumption became so embarrassing to pious Christians that the Church routinely passed "sumptuary laws" banning such extravagances. These were just as routinely ignored by the rich and some clergy, who pranced about in dazzling finery made all the more conspicuous by the contrast with the rough-sewn wool and coarse linen worn by nearly everyone else.

But these baubles had one positive side effect that would eventually alter the mind-set of Europeans as profoundly as the new thinking among certain scholars: the trade that delivered the goods. As more silks and perfumes were shipped in and raw goods such as grain and wool were shipped out, the nascent network of ships, shipyards, ports, accountants, merchants, sailors, and investors grew, filling the shipping lanes of the Mediterranean with Latin goods for the first time since the fall of Rome.

Soon this web of trade reached inland from the ports, giving rise to towns and cities along the highways into central Italy, France, and Germany—which in turn became bases of operations for merchants, muleteers, craftsmen, innkeepers, sheriffs, ne'er-do-wells, and financiers. The pace was most brisk in Italy, where merchandise arrived from Europe's interior to be loaded on ships in Venice, Naples, Pisa, and Rome and shipped to Byzantium and Syria. These vessels then returned to Italy with holds stuffed with wares to be offloaded and carried in caravans to Paris, Cologne, distant London, and hundreds of expanding market towns in between.

Ideas and information also arrived from afar, stimulating the minds of those Latins who met the oddly dressed merchants from the Moorish capital of Córdoba or from Arab-ruled Sicily and watched them use strange devices such as the astrolabe. Europeans also heard the strangers tell stories about faraway places—mostly yarns of the *Arabian Nights* variety, but also shop talk about numbers, bookkeeping, navigation by the stars, and how to design a better warehouse. This intercourse, though it affected only a tiny percentage of Latins, provided at least a peek into the East's advanced state of knowledge in fields such as mathematics and astronomy. A few intrepid Europeans even visited Sicily, Constantinople, Egypt, and Syria, accompanying trading vessels or, in the case of the Crusades, conquering so-called infidels.

Inevitably this tentative contact with far-flung cultures set certain Latins to scratching their heads over the issue of calendars and measuring time—not from a standpoint of theology, philosophy, or the endless *computus* tinkerings of monks, but rather from the practicality of having to draw up contracts with dates of delivery, inventories, and accounting records. This process was at least as important as the contribution of intellectuals in shifting the perception of time among ordinary Europeans.

But two points of confusion soon emerged among the practical-minded on the docks and in the markets, neither of which would be completely resolved for centuries: whose calendar, and whose number symbols and counting system should be used?

The first conundrum grew out of the multitude of methods, formal and informal, that people in this period employed to measure time. Arab merchants used Islam's lunar calendar and various versions of civil solar calendars, while Europeans continued to use Caesar's basic calender: 365¼ days, 12 months, 7-day weeks. Yet even in Europe details varied widely. For instance, no consensus existed on matters as basic as when the year started, which could vary from town to town and fief to fief. Some localities celebrated New Year's Day on Christmas, called *stylus nativitatis* (Christmas style) or *stylus curiae Romanae* (style of the Roman curia), since the papal chancellery sometimes opened their year on December 25. Many people used the date inaugurated by Caesar and used by the old empire: January 1, dubbed *stylus communis* (style of the people) and occasionally *stylus circumcisionis,* since this was the feast of the circumcision of Jesus. Other communities had the year starting on Good Friday, or the day after, or on Easter. Still others began their year in March, around the time of the vernal equinox, when some old German calendars and Rome's pre-Julian calendar began. Incredibly, this custom prevailed in Britain (and the American colonies) until 1752, when the Gregorian calendar was finally accepted by order of Parliament, and New Year's Day was moved from March 25 to the date everyone else in Europe had by then adopted: the first of January.

The naming of dates also varied as widely as ever. Many educated Latins still used the Roman kalends, nones, and ides, though more people were switching to our modern system of *dies mensis*, counting the days from 1 to 28, 29, 30 or 31. Other date reckoners used letters and syllables for naming the days. Most popular of all was the contin-

ued use of naming days for saints and feasts, despite the confusion of
different localities attaching their own saints to certain days. Even
widely celebrated holy days were sometimes observed on one day in,
say, Hamburg, and on another in Sussex.

These calendric differences were not a problem during the long
centuries when almost all communication and commerce had ceased.
When no one cared if it took weeks to get to Rome, and only the
occasional ship from Constantinople or Antioch docked in Venice, it
didn't matter if one was a day or two late, or if two different Christian
martyrs in two different localities were worshiped on the same day.
As trade grew more lively, however, people tried to sort out the Babel
of day names and dates—with little success. This is because no central
authority existed to standardize the calendar other than the Church.
St. Peter's though, remained firmly locked into the notion that time
belonged to God, not to bankers and sea captains: a core belief that
would have to be changed before the calendar could be reformed.

In 1100 the prospects for this happening seemed next to nil, even if
a few people were noticing that a great deal in nature and in commerce
seemed to operate with its own coherent set of rules independent of
church doctrine. Still, virtually all Europeans continued to believe that
God controlled everything and that truth was revealed to humans only
inasmuch as God allowed. So ingrained was this thinking that the
earliest conservative reaction to the new knowledge was not only
condemnation but dismay that anyone would waste time on such
wrongheaded notions as attempting to more accurately measure time.
One conservative writing in the mid-1100s assailed the ceaseless inquiries of certain scholars into "the composition of the globe, the
nature of the elements, the location of the stars, the nature of animals,
the violence of the wind, the life-processes of plants and roots."

A young Turk of the era, the French philosopher William of

Conches (1100–1154), responded with an outburst of support for objectivity that sounds like Roger Bacon a century later:

> Ignorant themselves of the forces of nature and wanting to have company in their ignorance, they don't want people to look into anything; they want us to believe like peasants and not to ask the reason behind things. . . . If they learn that anyone is so inquiring, they shout out that he is a heretic, placing more reliance on their monkish garb than on their wisdom.

Conches got away with such stridency in part because his argument remained obscure and his ideas outlandish to the mainstream. It would be another century before such new thinking became widespread enough that traditionalists would more actively try to thwart it. Besides, the sum total of the new knowledge remained modest in 1100, with scholars forced to search for answers in the few texts that had survived the dark years, many of them encyclopedic summaries of certain ancient works and ideas, but incomplete and often poorly written.

Yet even as scholars and would-be scholars despaired, a few pioneering thinkers from Europe were learning about and beginning to visit the great centers of Arab culture thriving just beyond their frontiers. What they saw and heard about stunned and shamed them as they realized the extent of their own ignorance; what one scholar called *Latinorum penuria*, the poverty of the Latins.

Even the most enlightened scholars of the time could not imagine the extent of their loss. Trapped behind their veil of darkness, the Latins had entirely missed the Gupta's flowering of mathematics and astronomy and knew nothing of Aryabhata, Brahmagupta, and other Indian scholars. Some over the decades had heard rumors of Islam's golden age, but few, if any, had ever heard the names of al-Khwarizmi, al-

Battani, or al–Biruni. Most Europeans were ignorant even of the Byzantines, beyond a few key ports and cities in Italy that had kept in furtive contact over the centuries.

In part this was understandable. Most outsiders were enemies, including at times the Byzantines, who continued to challenge the Lombards and other Westerners for control of southern Italy, and were sometime rivals in the east during various crusades. As for the Arabs, they stood like a colossus astride the borders of Europe, a military superpower that fearful Christians thought of not as an enlightened culture of scholars but as the army of Satan himself. How else to explain their triumphs against God's people?

In a whirlwind they had snatched away Spain and stormed across the Pyrenees to seriously threaten France. Conquering large chunks of the old Roman Empire, including all of North Africa, they had launched raids throughout the 800s from the Mediterranean into France and Italy. In 827 they captured Sicily and in 838 their armies fought at Naples, summoned by Lombards as allies against the Byzantines. Four years later the Arabs garrisoned a base at Bari, on Italy's heel. Four years after that, in 846, Arab squadrons landed at Ostia and threatened Rome. Unable to penetrate its walls, they sacked the Vatican cathedrals of St. Peter and St. Paul, which lay outside the main city walls, and desecrated tombs of the popes.

During the 900s they launched raids from Italy and Spain deep into central Europe, capturing towns that still bear Arab names as far north as Switzerland. For three centuries, from the mid-700s to the mid-1000s, Arab armies and raiding parties menaced the western Mediterranean, dominating its sea lanes and leaving Europeans terrified they would launch a major invasion.

But the Arabs brought far more to their new domains in Europe than curved sabers and copies of the Koran. Following the pattern of their earlier invasions in Asia and Africa, the era of conquest in Spain and Sicily soon gave way to periods of cultural assimilation and learning. Against the backdrop of raids and skirmishes on the frontier, art and scholarship flourished in the new Moslem cities, where scholars gathered under the patronage of caliphs and emirs who imported vast numbers of texts to fill new libraries built in Córdoba, Seville, Toledo, and Palermo. This rush of knowledge finally brought to the frontiers of Latin Europe works by ancient Greeks, Romans, and Indians, and the latest works by Arabs writing on everything from the anatomy of the human eye to Hindu numbers.

In Spain, al-Khwarizmi's *Algoritmi de numero Indorum* and other texts had reached Córdoba by the late ninth century, joining a vast treasure trove of manuscripts housed in a new library built by Caliph Abd ar-Rahman III (891–961)—a patron of art and learning who filled Córdoba with monumental buildings that fused Arab, Romanesque, and Persian motifs in the style known as Moorish, with its graceful arches, fluted columns, onion domes, and vast gardens. Under his successors, the collection of texts begun by Abd ar-Rahman was said to house 400,000 volumes, which if true meant it rivaled the number of volumes in Alexandria's library.

Likewise, the emirs governing Sicily imported texts and encouraged learning, though the island's apex of Arab culture came not under their rule but after it was conquered by a Christian: Roger Guiscard (1031–1101), son of a baron in Normandy.

Originally a mercenary seeking riches and adventure, Roger trekked to southern Italy from France in the 1060s to join four of his brothers in the long-simmering struggle over this disputed territory—which they ended by throwing out the Lombards and Byzantines and seizing the place for themselves.* With Roger in the lead, they also invaded Sicily, tossing out the Arabs by 1072.

Once secure in his capital of Palermo, Roger renamed himself Roger I, count of Sicily. He then established one of the odder amalgamations of cultures in the Middle Ages, blending Christian and Moslem with older currents of history on an island rich in Greek, Roman, and Byzantine traditions. By Arab terms an uncultured Christian, Roger nonetheless won the loyalty and admiration of Moslems,

---

*Roger was one of twelve brothers.

whom he welcomed into his domain—including Arab soldiers and advisors, and a stable of Eastern scholars, philosophers, and astrologers.

Two of Roger's successors expanded on this strange Arab–Norman cocktail. His son Roger II (1095–1154), known as the "half-heathen king," ruled both Sicily and southern Italy like an Arab sultan, dressing in Persian silks and opening up his court to Moslem intellectuals. Roger II's grandson and successor, Frederick II (1194–1250), inherited not only Sicily and southern Italy but also Germany and the crusader kingdom of Jerusalem. Elected Holy Roman Emperor in 1220 Frederick maintained an oriental-style harem and surrounded himself with philosophers and sages from Baghdad and Syria, dancing girls from the Orient, and Jewish scholars. From Syria he imported experts on falconry; from Spain he brought a translator who created a Latin summary of Aristotle's biological and zoological works. Frederick founded the University of Naples in 1224, endowing it with a large collection of Arabic manuscripts on Aristotle and other ancients. Copies of Latin translations were sent to the universities in Paris and Bologna. Frederick also led a successful Crusade to Palestine in 1228–1229—the fifth crusade—and recaptured Jerusalem, Bethlehem, and Nazareth.

Still, the infusion of Arabic knowledge was very slow, with only a few scattered documents making the journey from Córdoba, Palermo, and Damascus before 1200. Some of the earliest translations were penned in northern Spain beginning in the mid-tenth century at the monastery of Santa María de Ripoll at the foot of the Pyrenees, mostly works on geometry and astronomical instruments. Next came works by Plato, Euclid, Aristotle, and others, coming out of Roger's Sicily, northern Spain after the fall of Toledo to the Christians in 1085, and Byzantium and Palestine as the Crusaders stormed across the east starting in 1096.

Leading translators and collectors of manuscripts in these early days included Gerbert of Aurillac (c. 946–1003), later Pope Sylvester II. He trekked to northern Spain to carry home Latin translations of Arab treatises on the abacus and the astrolabe. Another was Adelard of Bath (c. 1075–1160). He journeyed by ship along the new eastern trade routes to the Crusader-held coast of Syria, where he translated Euclid into Latin using Arabic translations of the original Greek. Most prolific of all these early translators was the Italian Gerard of Cremona (c. 1114–1187). Fluent in Greek and Arabic, he was a leading figure in the new college of translators set up by the Spanish archbishop Raymond after the capture of Toledo (and its library), rendering into Latin texts by Galen, Aristotle, Euclid, al-Khwarizmi, and Ptolemy, among many others.

Independent-minded thinkers in Europe welcomed each precious manuscript with fascination, though the transfer of knowledge was hardly swift or comprehensive. Most Europeans, even those with some education, remained locked in the timelessness of the Middle Ages and remained ignorant of the new knowledge. Others condemned the texts as products of pagans and devils. Still others resisted anything new because they either failed to understand it or preferred their own familiar ways and traditions—much as Americans today continue to use inches instead of centimeters. Even those who embraced Aristotle and al-Khwarizmi were often confused by poor translations and by the random selections that arrived: a fragment of a Platonic dialogue one year and a chapter or two of Euclid the next.

An example is the reception of the new Hindu numbers as they completed their journey west from India to the Arabs and onward to the Latins. Indeed, the Europeans took centuries to fully integrate what the Arabs had largely absorbed just a generation or so after the 789 arrival of Kanaka in Baghdad.

The first Hindu number known to have been scrawled on a European manuscript appeared in northern Spain in 976 and used the "western" Arabic form of the numbers one through nine.

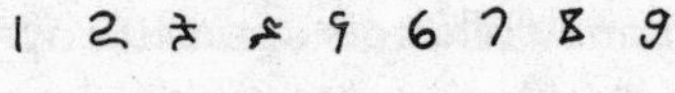

Twenty years later, in the 990s, Gerbert of Aurillac taught the Hindu numbers to his students, undoubtedly picking them up after a stint in Spain. But Gerbert apparently failed to understand their computive power and limited his use of them to special counting boards. These boards failed to catch on, however, in part because people who tried to use the boards had no idea which way was up for the strange symbols. For instance, they seem to have confused a  with a .

Mention of the numbers all but disappeared for another entire century until the Englishman Robert of Chester (c. 1100) visited Spain and translated al-Khwarizmi's little book into Latin in 1120. This and other translations of al-Khwarizmi inspired several Latin textbooks on the "new arithmetic," including descriptions of the decimal system and positional notation. Still, it took several more centuries before Europeans entirely abandoned Roman numerals, despite their clumsiness and inferiority to Hindu-Arabic numerals. Even bankers and merchants resisted them at first, worrying that they were easier to falsify than Roman numerals. Some less-educated merchants also suspected that the symbols were a secret code used by orientals and other Europeans to cheat them.

As late as the fifteenth century, when Hindu-Arabic numbers took the form we now use, Europeans were still having trouble making the transition. In a preface to a calendar in 1430 the maker gives the length of the year as "ccc and sixty days and 5 and sex odde howres." Later in the century, two years after Christopher Columbus sailed to America, another author described the year as MCCCC94—1494. Yet another used the new positional system along with the more recently arrived zero, but mixed up Hindu and Latin numbers to come up with the year, 1502, written as IV0II, with I (1) in the thousands place, V (5) in the hundreds, 0 (zero) in the tens, and II (2) in the digits. Dutch painter Dirck Bouts (c. 1400–1475) dated a painting he placed in the cathedral in Louvain with MCCCC4XVII. This may be 1447, but who knows?

Progress was equally slow for other mathematic concepts crucial to fixing the calendar, including decimals and zero, neither of which was routinely taught in universities until at least the mid-fourteenth century. The first systematic treatment of decimal fractions in Europe had to wait until 1582, the year of the Gregorian calendar reform, when Dutch mathematician Simon Stevin (1548–1620) explained the system in a book called *La Thiende* (The tenth). But Stevin did not use our modern form for his decimals, having no decimal point. He would have written the fraction for the length of the solar year as:

$$\overset{\textstyle ⓪①②③④⑤}{3\ 6\ 5\ 2\ 4\ 2\ 1\ 9\ 9}$$

instead of:

$$365.242199$$

The invention of the decimal point is usually attributed to either mapmaker and Galileo rival G. A. Magini (1555–1617) in a 1592 work, or to the leading astronomer on Gregory XIII's calendar commission, Christopher Clavius (1537–1612), who used them in a table of sines in 1593.

As for zero, its first significant appearance in Europe comes during the eleventh and twelfth centuries, at roughly the same time as the other nine Hindu-Arabic numbers started to come into wide use—first as a place marker on counting boards devised by Gerbert and others, then as a digit in positional notation. It took longer for zero to be thought of as a real number in mathematical equations, though by the turn of the seventeenth century it and positional notation were familiar enough for William Shakespeare to use them as a metaphor for infinite gratitude in *The Winter's Tale,* written in 1610:

Like a cypher . . .
I multiply with one, "We thank you,"
Many thousands more, that go before it.

This reticence over something as basic as numbers begins to explain why it took so long to reform the calendar, a process that was far more difficult and complicated than deciding whether to use 5 instead of V, or 365 instead of CCCLXV. For unlike numbers—or zero or a decimal fraction—the calendar belonged to God, and was assumed to be an immutable timetable of faith and worship that no one had dared challenge, not even the likes of Bede and Hermann the Lame. Which made the entire question of time and the calendar more and more perplexing as Europe reawakened and time ceased to be something that one could ignore or relegate exclusively to God.

Whether traditionalists liked it or not, secular time was restarting in Europe, and with it a need to reevaluate the nature of time—how to measure it, use it, and understand it. This issue lay at the heart of a looming larger question: how to react to an influx of new ideas that in some cases directly challenged not only details of Church dogma but the fundamental beliefs of an entire society. This would become the central dilemma of scholars from 1100 to 1300: how to account for knowledge that seemed to come out of nowhere, and in essence offered a new kind of religion that put its faith in observation and logic. It was this debate that would ring across Europe during the High Middle Ages, primarily in the halls and courtyards of this era's profound new invention: the university.

11

The Battle Over Time

Since the General Council forbade any alterations in the calendar, modern scholars have had to tolerate . . . errors ever since.
—JOHN OF SACROBOSCO, 1235

Imagine the 14-year-old son of a wealthy shipowner in Pisa, or the second son of a prosperous squire in Kent, circa 1240. How would each have reacted to the news that his father, with the support of the local lord, was sending him to a university in Bologna, or in Oxford?

They might have been dimly aware of the new knowledge arriving in Europe, particularly the boy in Pisa. In the harbor he would have seen dark-skinned Arab traders in turbans haggling with his father and scratching out strange, compact symbols for numbers that differed from the ones Latins used. He might also have heard from former university students about the halls at Bologna, where black-robed masters delivered lectures revealing untold secrets of the ancients: powerful knowledge that his father wanted him to learn so that he could help the family. But it was also dangerous knowledge, or so he might have been told by a local priest or elder looking out for the boy's spiritual well-being, who warned him to beware of ideas that would offend God and the Church.

The boys in Kent and Pisa would have left home in the early autumn, sometime before lectures began on or just after St. Michael's day, September 29, or some other date no one would have been precise about following. Beyond this, the boys probably gave no more thought to dates and exact times than our farmer on the Rhine or the weaver in France did in the year 800. By now a few people were using reasonably accurate water clocks, but mechanical clocks had not yet been invented—at least, there are no definitive records of any. And public bell towers clanging each hour from the town square remained decades in the future, with several rising up over cities near Pisa in the early and mid-1300s and the first large clock appearing in England at Windsor Palace in 1351.

Otherwise, the boys would have reckoned time by looking up into the sky, eyeballing the arc of the sun as Geoffrey Chaucer does in *The Canterbury Tales* to move us through the timeline of his journey. As the day in his story ends, Chaucer writes (in updated English):

> *From the south line the sun had now descended*
> *So low, it stood—so far as I had in sight—*
> *At less than twenty-nine degrees in height.*
> *Four o' the clock it was to make a guess;*
> *Eleven foot long, or little more or less,*
> *My shadow was, as at that time and place,*
> *Measuring feet by taking in this case*
> *My height as six, divided in like pattern*
> *Proportionally; and the power of Saturn*
> *Began to rise with Libra just as we*
> *approached a little thorpe.*

Chaucer, who also wrote a treatise on the astrolabe, was undoubtedly more adept at making such "guesses" than our young men from Pisa and Kent. Yet his inclusion in his tales of references to angles of the sun suggests his audience by the mid-1300s was familiar with the idea, though even then the times and measurements are given as "more or less," as if this is close enough for pilgrims trekking leisurely along the road to England's most holy city.

Packing up a satchel on the appointed day, the squire's son in Kent would have started off with prayers for a safe journey in the cool darkness of his village church. Then, before the sun rose too high, he would have set off for London, possibly accompanied by a servant, joining a highway like the one Chaucer wrote about: filled with messengers, knights, monks, merchants, ne'er-do-wells, highwaymen, and pilgrims.

London would have seemed enormous to this country boy. A city of perhaps 20,000 packed inside thick stone walls, it drew in people from all over England, and a few from foreign lands, with ships at anchor on the Thames from as far away as the Levant. Merchants bought and sold in markets reeking of dung, perfume, and exotic spices. Our student-to-be would have seen beggars in rags wailing for a few grains of barley, courtiers in colorful livery from the royal palace, soldiers sporting broadswords in sheaths attached to their belts, and merchants from France and Italy calculating everything so quickly on abacuses that one could hardly see their fingers move.

Staying overnight in a London inn, the boy would have continued eastward toward Oxford, following the snaking, narrowing, lazy flow of the Thames, and passing by hedgerows ablaze with autumn colors and the earthy smell of fields turned over for the winter. Arriving at the gates of Oxford, the squire's son would have seen a small, sleepy market town along the river, where perhaps a few hundred students had come to live among the townspeople—who often found the students loud and obnoxious. Oxford at one point shut down between 1209 and 1214 when a student killed a townswoman, and a local mob hanged two or three students in retaliation.

Mostly the students stayed in modest houses of stone and thatch in the neighborhood surrounding St. Mary's Cathedral. The boy would have seen none of today's grand quads, libraries, and other university buildings, because in 1240 they had not yet been built. The only evidence that he was in a university town was the sight of other boys and men dressed in black robes; among them a scattering of masters, including perhaps Roger Bacon, who might have been teaching then at Oxford.

Taking a deep breath, the boy would have turned into one of the small, cramped buildings where he was told the registrar kept his records, just as his counterpart in Bologna was strolling into the equivalent office to matriculate in this ancient city in northern Italy. Neither realized that he was headed toward an encounter with the unknown unlike anything their fathers or grandfathers could have imagined, a new way of approaching the world that was already turning the university into a major intellectual battleground between the forces of faith and reason, the sacred and the secular. This battle would forever alter the way Europeans thought about themselves and the universe, and it would shift the fundamental perception of time away from "more or less" to ever more precise expectations in the generations that followed the boys from Kent and Pisa.

The universities did not start out as crucibles for an intellectual revolution. Originally referred to as *universitas magistrum* or *universitas scholarium*—"university of masters" "university of students"—at first they were little more than gatherings of students in certain cities, attracted by masters whose fame allowed them to charge fees. Many of the earliest university teachers came from the ranks of translators who had trekked to Toledo and Sicily and returned to teach the "secrets" of Aristotle, al-Biruni, and Euclid. These universities operated in rented halls and hostels, with wealthier students renting their own rooms, and those with fewer resources, such as our shipowner's son and squire's son living either with their masters or in inns and local hostels. The spirit of these enclaves was one of a shared adventure in learning—a profound experience for young men coming from Pisa, Kent, and elsewhere. Students with more ecclesiastic leanings, or in economic straits and in need of free housing, joined one of the new Catholic orders attached to the universities, such as the Franciscans and Dominicans.

The greatest of the early masters was Peter Abelard (c. 1079–1144), son of a minor Breton lord and a proponent of the newfangled logic of Aristotle. An intoxicating lecturer, he is sometimes credited with single-handedly attracting the original crowds of students that made possible the university in Paris. The young Abelard epitomized the sort of person drawn to the new style of learning in the twelfth century. Brilliant and relentless in his scholarship, freewheeling and passionate in his lifestyle and personality, he represented a profound shift away from the cloistered approach of learning and toward a search for the truth in open discourse and disputation, and through the unfettered power of his intellect.

Predictably, conservatives criticized the new thinking and the entire project of the universities, launching a centuries-long battle between traditionalists and men such as Abelard. As early as the 1060s a leading cardinal, Pier Damiani (1007–1072), warned that the new learning represented a grave danger to bedrock medieval beliefs and might eventually cause a split between the world of reason and that of faith. He and others of a similarly contemplative bent worried not only about offending God but about the unsettling effect on the faithful should basic tenets of the Church be undermined. Less philosophical critics simply condemned the new teachings, calling heresy anything that contradicted the Church. Still, the universities proliferated, with Bologna the first to receive an official charter in 1088. Paris received its in 1150 and Oxford in 1167, though the rush did not come until the 1200s and 1300s, when dozens of schools were officially opened, from Salamanca in Spain (1218) to Krakow in Poland (1364).

The university curriculum began with training in four or five general areas: theology, law, medicine, arts or philosophy, and music. The masters also taught what was known about astronomy, mathematics, and other sciences, though these more empirical subjects tended to be overshadowed by the deep philosophical and theological controversy touched upon by Hermann the Lame, promulgated by the Arabs, and shouted about by Abelard: what to do about the growing evidence that two truths existed, that of the Church and that suggested by nature and reason.

This was hardly a new quandary. It revisited an old debate from the waning days of the Roman Empire, depicted by St. Augustine as the "city of God" versus "the city of man." It also was a recasting of the ancient dispute between, on one hand, the Aristotelian notion of the particular and the individual, of empiricism and logic, and, on the other, the Platonic ideal that the general and the universal are everything, and that perfection exists but is beyond human comprehension. In ancient times a great pendulum had swung back and forth between these two worldviews, with Caesar and the Rome of the early emperors representing a swing toward the secular, and Constantine and later Augustine swerving over to embrace the sacred.

Now for Europe in the High Middle Ages, this debate had returned in full fury to become an epochal argument, one that would either propel it into a new age of empiricism and secularism or sustain it in a world of mysticism and faith.

For several centuries, until long after Copernicus and even Galileo, the outcome would remain unclear, with traditionalists fighting back at every turn. Abelard himself was eventually destroyed, in part because of his own outrageous departure from acceptable behavior when he wooed the young Héloïse, the teenage niece of a prominent canon in Paris, had a son with her, then married her in secret—which prompted the girl's irate uncle to have the master scholar castrated. More serious for Abelard's career, if not his anatomy, was his practice of intentionally upsetting his enemies by publishing such works as his *Sic et Non,* which explicitly laid out the contradictions of various Church leaders on important theological points. He also challenged orthodox views on the nature of God, Christ, and the Holy Spirit, a sore spot of Catholicism since the age of Constantine, when the Council of Nicaea condemned Arianism over this same issue.

After being charged with heresy for his ideas, Abelard retired to a hermitage, became an abbot, and was eventually tried by his enemies—led by Bernard of Clairvaux (1090–1153). The French-born leader of a movement toward more mysticism and reliance on faith, not less, Bernard spoke for the old guard when he criticized those who learned "merely in order that they may know," insisting that "such curiosity . . . is blamable." Calling Abelard a "hydra of wickedness" he condemned all learning that was not directly necessary to serve God, proclaiming that the only road to truth was to maintain a "pure conscience and unfeigning faith."

Abelard's downfall did not squelch the new thinking, as Bernard undoubtedly hoped it would. But it did remind scholars of a need to be prudent in what they said and wrote, at least in public. "When the object of the dispute can be explained more clearly through the rules of the art of logic," wrote the Italian scholar and ecclesiastic Lanfranc (c. 1005–1089), a confidant of William the Conqueror and later the archbishop of Canterbury, "I conceal the logical rules as much as I can within the formulas of faith, because I do not wish to seem to place more trust in this art than in the truth and authority of the Holy Fathers."

Despite this, a growing number of intellectuals followed Abelard's lead in seeking truth through logic and nature—though few as effectively as an Arab in Córdoba named Abu al-Walid Mohammed Ibn Rushd (1126–1198), known in the West as Averroës. A teenager when Abelard died, Ibn Rushd lived in an era when the Islamic world itself had been locked in a debate between the sacred and the secular, with the same enormous stakes. By now, however, the great Islamic empire was long gone, succeeded by shifting emirates and sultanates that tended to be religiously conservative and uninterested in learning. And with it was gone the era of the House of Wisdom, when Aristotle and Mohammed could be studied side by side. One exception had long been Moslem Spain, though it was now ruled by North Africans more orthodox than previous emirs, even as the Moors slowly lost territory to the Christians.

It was against this backdrop that the physician, judge, and philosopher Ibn Rushd wrote what Europeans considered the most thorough and enlightening commentaries to date on Aristotle and the Aristotelian universe. Called "the commentator," a play on Aristotle's appellation "the philosopher," Ibn Rushd conceived of a philosophical argument that tried to solve the dilemma between the sacred and the secular by insisting that two contradictory truths could exist: one for science and "natural reason" and one for "revelation." According to his philosophy:

> When a conflict arises we will therefore simply say: here are the conclusions
> to which my reason as a philosopher leads me, but because God cannot lie,
> I adhere to the truth he has revealed to us and I cling to it through faith.

At first Ibn Rushd's "double truth" was merely frowned upon. Then it was aggressively challenged by religious authorities in Christian Europe and Islamic Córdoba. Declaring that Aristotle was not a god, but a man and therefore fallible, bishops and imams alike objected to Ibn Rushd's insistence that science was on par with divine truth. They also were horrified by Ibn Rushd's assertion that while science proved that God was the mechanistic mover of the universe, God himself was a "machine" entirely removed from interference in human affairs. According to Ibn Rushd, it was the laws of nature—of this machine—that uphold the eternity of the universe and the passage of time. This idea denied a range of core Christian and Moslem beliefs, including creation, the doctrine of an active and fully engaged God, and the immortality of the individual soul.

Ibn Rushd's ideas nonetheless resonated with many intellectuals in

Europe, working their way into a gradual rethinking of time by Christians, begun with the likes of Hermann the Lame. For instance, around 1200, a Norman mathematician and encyclopedist named Alexander of Villedieu suggested there may be two truths in regard to time reckoning. He makes no direct mention of Ibn Rushd's work, and as a pious Catholic he would have been horrified to be mentioned in the same sentence with this near-heretical Arab. Yet he was advanced enough in his thinking to use Hindu numbers, and was reasoning along the same line as Ibn Rushd when he divided the measurement of time into two categories: what he called philosophical *computus,* by which he meant time as measured by science, which is infallible; and ecclesiastic time, which he curiously referred to as "vulgar" *computus,* "the science of dividing time according to the custom of the Church." But Alexander dodges the potential controversy of his categories by telling us that he does not want to discuss the philosophical *computus,* but will confine his comments to the ecclesiastic.

Eventually the Italian master Thomas Aquinas (1225–1274) solved the dilemma, at least temporarily, by rejecting the incompatibility of the two truths. He argued that in fact both "truths" point in the same direction: toward God and toward the universe of ideas and morals created by God. To do this Aquinas made the breathtakingly bold assertion that Platonic universals could be proven by Aristotelian logic. In other words, this brilliant Italian philosopher and theologian, born in a castle to the noble counts of Aquino and trained in Naples and Cologne, attempted in a comprehensive manner to unite the worlds of Aristotle and Plato.

Part of Thomas's argument rested on a theory that time and the universe could not be eternal, as Aristotle claimed, but must have started with an original, unmoved mover, which Thomas says is God. He then sets out in his massive *Summa Theologica,* which he worked on until his death in 1274, to apply the rules of science as argued by Aristotle to prove the reality of God's perfection, of the Creation and the existence of the human soul, and of the ethical foundation of Christian virtue. This attempted conciliation of the sacred and the secular provided the great philosophical compromise of the Middle Ages, permitting intellectuals on both sides of the great divide of the two truths some breathing room.

But Thomas's opus was not initially well received either by the followers of Ibn Rushd, who accused him of faulty logic, or by the Church. At first conservative Church leaders condemned his *Summa* as being overly radical, though just a generation after Thomas's death his philosophy was embraced by the Church. It became the official theological response to the new knowledge and a counter to Ibn Rushd—a point vividly made in a painting rendered during this period of an enormous Thomas enthroned, "crushing" under his feet a tiny, bearded, and turbaned Ibn Rushd. Thomas was made a saint in 1323.

For a time Thomas's philosophy comforted conservatives and scholars who shared Alexander of Villedieu's discomfort in acknowledging truths that seemed to contradict the Church. But it also gave a green light of sorts for science to seek its own truths, though within strict limits—as Bacon would discover, and many years later Galileo. Another period painting amply demonstrates this, illustrating a gigantic St. Augustine, dressed in glittering medieval robes, crushing underfoot a tiny Aristotle in a simple tunic. Yet the Aquinas compromise had the advantage of at least quieting the all-encompassing theological debate so that men such as Bacon could begin to turn their attention toward using the new knowledge of the Greeks, Arabs, and Indians for scientific endeavors, rather than to score points in heated philosophical debates.

But Ibn Rushd and like-minded Islamic scholars did not win even this partial victory in their own homeland. Toward the end of Ibn Rushd's life the conservatives in Spain struck hard against the celebrated schools in Córdoba, denouncing Ibn Rushd and other intellectuals and later disavowed his work. For even as Europe finally began to absorb the learning brought to its frontiers by Arabs, the world of Islam was falling deeper into a period of political turmoil and outside threats from Mongols and others, hastening a growing chill in its intellectual life.

With the fate of the human soul and the beliefs of a thousand years hanging in the balance, scientific pursuits remained largely fringe endeavors for intellectuals during the 1100s and 1200s. Peter Abelard, for one, brushed off mathematics, astronomy, and virtually all science, insisting in 1140 that "philosophy can do more than nature." As for time reckoning, he dismissed it as being in the same low category as usury, useful in the hallowed halls of the university only for collecting fees from students based on elapsed time. Thomas Aquinas a century later was equally dismissive of time reckoning, refusing to allow that it was real in Aristotelian terms. Like Abelard, Thomas argued that time fixing should be excluded from the theoretical sciences, also ranking it as a lowly mechanical art unworthy of scholarly contemplation. Even those who pondered the new texts with an eye toward learning more about science tended to simply read Ptolemy, Galen, Euclid, and the Arab astronomers, rather than trying to apply their ancient ideas to anything new.

Still, a scattered handful of scholars pored over the mass of new knowledge and tried to make sense of it, and attempted to apply it to everything from human anatomy to more accurately measuring time.

One of the first hands-on time reckoners steeped in the new knowledge was Reiner of Paderborn (c. mid-twelfth century), dean of the cathedral at Paderborn in the northern Rhine Valley. Now all but forgotten, Reiner wrote a treatise in 1171, *Computus emendatus,* that applies the new Hindu numbers and mathematics to the old formulas of *computus* involving the Easter calculation—and proves that the old 19-year lunisolar cycle was misaligned with the true movements of the sun and moon. This error amounted to one day lost every 315 years—that is, every 315 years the 19-year cycle of lunar and solar years slipped a day against the Julian calendar. Reiner's measurements also led him to the near-heretical conclusion that all attempts by computists to date the age of the world, and to create a timeline of history dating back to creation, were mistaken, given the errors in the calendar.

In 1200 Conrad of Strasbourg wrote that the winter solstice had fallen behind by 10 days since Caesar's time. Conrad's estimate established the figure of 10 days as gospel among reform-minded time reckoners, though they argued about whether one should calculate the drift from Caesar's founding of the calendar in 45 B.C. or from the time of the Council of Nicaea in 325, when time reckoners fixed the equinox on March 21.

A few years after Conrad, the English scholar Robert Grosseteste (c. 1175–1253) recalculated Reiner's lunar-solar slip and amended it to the gain of a day every 304 years—closer to the actual drift against the Julian year of a day every 308.5 years. He also proposed a solution: that one day be dropped from the lunar calendar every three centuries. Grosseteste, chancellor of Oxford University and later a bishop, also closely studied measurements of the solar year, confirming once and for all that the values arrived at by Hipparchus, Ptolemy, al-Battani, and other Arabs and Greeks were superior to those worked out by Bede and centuries of computists. This led him to suggest a new starting point for the Easter calculation—a spring equinox of March 14 instead of March 21—to compensate for the centuries-long drift in the calendar against the calendar year. Grosseteste is also remembered because of the standards he set for science. Known for his work in geometry and optics as well as astronomy, he was an early advocate of using experimentation and observation to verify theories. This was an idea years ahead of its time. For while most intellectuals were trying to reconcile contradictions between the new knowledge and the old dogma, Grosseteste was taking the next step and trying to reconcile the contradictions between reason and experience—between the new knowledge as written in books and empirical evidence.

By Grosseteste's time few serious time reckoners were denying that the errors existed in the lunar and solar calendars. But this hardly meant they were all for reform. Another Englishman, John of Sacrobosco (c. 1195–1256), proved the errors down to the minutes and seconds using an astrolabe and a deep knowledge of Arab, Greek, and Indian mathematics and astronomy. Yet he was able to offer only one modest reform in the solar calendar: that the calendric order be restored by canceling the leap day every 288 years. Otherwise John stuck with Bede's admonition to follow the "universal custom" of accepting the errors, insisting that the Church was the final authority. Referring to the Council of Nicaea in 325, he wrote: "Since the General Council forbade any alterations to the calendar, modern scholars have had to tolerate errors ever since." John's reticence must have resonated with scholars. For three hundred years his textbook on time reckoning remained a standard in universities. Even Protestants republished it in 1538, soon after they changed the university at Wittenberg to a Lutheran institution.

Into this mix in the mid-1200s came Roger Bacon, another firebrand visionary along the lines of Abelard. He not only took up the cause of Robert Grosseteste in pushing for reform of the calendar but he also became a staunch advocate of Grosseteste's championing of empiricism and the objectivity of science. Even further ahead of his time than Grosseteste—centuries further—Bacon demanded that scholars stop talking and debating and start *doing*. In his *Opus Maius*—written in the 1260s, the same decade that Thomas Aquinas was laboring over his *Summa Theologica*—Bacon writes:

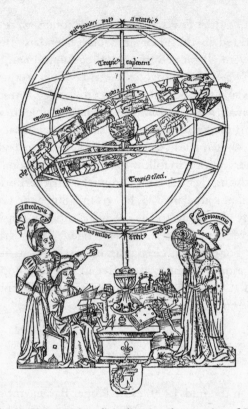

The Ptolemaic Universe; Astroligia directing the attention of Sacrobosco to Ptolemy, from *Urania Ptolemaeus,* 1538.

The Latins have laid the foundations of knowledge regarding languages, mathematics, and perspective; I want now to turn to the foundations provided by experimental science, for without experience one cannot know anything fully.

In possibly Bacon's most famous passage he vividly illustrates his point:

If someone who has never seen fire proves through reasoning that fire burns, changes things and destroys them, the mind of his listener will not be satisfied with that, and will not avoid fire before he has placed his hand or something combustible on the fire, to prove through his experience what his reasoning had taught him. But once it has had the experience of combustion the mind is assured and rests in the light of truth. This reasoning is not enough—one needs experience.

As passionate and arrogant about his cause as Abelard was about the use of logic a century and a half earlier, Bacon argued that nature had been established by God and therefore needed to be explored, tested, and absorbed to bring people closer to God. He warns that a failure to embrace science is an affront to God and an embarrassment to Christians, who were forced to acknowledge the superiority of Arab science.

A prime example of this embarrassment, he said, was the habit of Christian time reckoners and mathematicians to round off numbers rather than trying to calculate them precisely. This was an intentional jab at time reckoners such as Bede and Bacon's contemporary John of Sacrobosco—those who admitted to calendric errors but settled for approximations rather than challenge the Church. This had led, writes Bacon, to a calendar that in that very year (1267) was causing havoc for devout Christians.

The errors I have mentioned are terrible in themselves, yet they bear no comparison to those which follow from the facts now stated. For the whole order of Church solemnities is thrown into confusion by errors of this kind respecting the beginning of the lunation according to the Calendar, as well as by the error in determining the equinoxes. And not to refer to other years for evidence of this error, I shall state the case in this present year.

Which he does in detail, explaining what this meant for pious Christians, in terms much starker than Reiner of Paderborn or Robert Grosseteste would have dared:

> Wherefore the feast of Easter, by which the world is saved, will not be celebrated at its proper time, but there is fasting this year through the whole true week of Easter. For the fast continues eight days longer than it should. There follows then another disadvantage that the fast of Lent began eight days too late; therefore Christians were eating meats in the true Lent for eight days, which is absurd. And again then neither the Rogations nor the Ascension nor Pentecost are kept this year at their proper times. And as it happens in this year 1267, so will it happen the year following.

Bacon's ardor for correcting obvious errors came in part from a belief that the Antichrist was about to arrive on earth; shortly after this event would come the end of the world. This left Christians little time to use science to bring order to civic life and to perfect the Christian way of life, or so Bacon argued in his strange amalgam of science and spiritualism.

Bacon did not stop with condemnations, however. He demanded a change—taking his case directly to the papacy when given his surprise opening in 1265 from Guy Le Gros Foulques, the soon-to-be Pope Clement IV.

But Bacon was not after a simple mechanical solution. He framed his argument philosophically by dividing time into three categories: that which is "designated by nature, . . . by authority, and . . . by custom and caprice." He defined natural time as the measurable passage of years, seasons, months, and days; authoritative time as that used in civil and ecclesiastic calendars; and time by custom as when people arbitrarily impose periods of time, such as months that number 28, 29, 30, or 31 days.

Bacon derived his three-part definition from Bede, though the venerable monk had concluded that the authority of God's time superseded the others. Bacon argued the opposite: that natural time *was*

God's time, and that time as interpreted by an authority such as the Church can be mistaken. This provided the philosophical underpinning that in Bacon's view gave Rome the right and the responsibility to correct the calendar, both as Europe's only authority capable of ordering a change and as God's authority on earth.

But Rome was still grappling with what to do about Abelard and the onslaught of reason, and it hardly seemed prepared to move several steps ahead to embrace Bacon's essential idea—that human intellect through experimentation and observation could correct and negate core Church teachings. This leaves us with a mystery as to whether or not Clement IV shared Bacon's ideas—or would have been receptive to his demand for reform had he lived.

One thing is sure: Clement's advisors, those who received and presumably glanced at the friar's opus, were *not* receptive. After the pontiff's death no one at St. Peter's so much as mentioned Bacon. Years later the French lawyer and bishop Guillaume Durand (c. 1230–1296), who joined the papal service under Clement, wrote an entire volume on time reckoning, without so much as mentioning Roger Bacon.

Yet change was in the air as the year 1300 arrived, much of it coming not from the endless debates and scholasticism of the universities or the gilded basilicas of Rome but more than ever from the rising merchants, traders, bankers, kings, generals, shipowners, and other practical-minded people. They who felt a keen need not only to measure time accurately but also to find better ways to build ships, plant millet, fashion swords, and construct battlements. By the 1290s the word *computus* itself had shifted its meaning to something more familiar to us today. Indeed, back in 1250 the Italian *conto* still meant time reckoning, while just a generation later a young Dante Alighieri (1265–1321) was writing love poems in which *conto* describes the relationship between two lovers—not physically, but in terms of economics and accounting, that is, how lovers reckon and balance income and expenditure. The word *computare* was becoming closely connected with finance, with *conto* in Italian, *cuenta* in Spanish, and later *kanto* in German, all meaning to count or reckon not stars and epacts of lunar cycles but money.

At the same time a civic calendar of sorts was beginning to take shape along with the renewed sense of linear time crucial to conducting ongoing business and government. This new age was signaled by Pope Boniface VIII's declaring 1300 to be a "century" year, which he marked with a jubilee to celebrate thirteen centuries of Christendom—starting a tradition that continues to the present, with jubilees now held every quarter century. Boniface's affair attracted 20,000 pilgrims drawn by the spectacle and by the pope's offer of special indulgences. The event also signaled a new awareness of the calendar, and the final triumph of Dionysius Exiguus's system of counting years from the supposed birth of Christ.

It also was an effort by Boniface to emphasize the primacy of Rome in an age when the papacy was being challenged by the power and authority of kings, dukes, and counts, and by what would become modern state governments replete with ministers, lawyers, tax and spending authorities, and bureaucrats. The jubilee recalled the lavish feasts held by ancient Roman emperors on special anniversaries, and was meant to demonstrate the supremacy of Rome and papal rule then and forever. This sentiment was made more starkly clear two years later when Boniface issued a papal bull that ordered all Christians to recognize the supremacy of the pope in all matters. Directed primarily at Boniface's enemies of the moment, Edward I of England and Philip IV of France, the pope wrote: "Therefore we declare, state, define and pronounce that for every human creature to be subject to the Roman pope is altogether necessary for salvation."

It says a great deal about change in Europe that a pope even needed to issue such a proclamation just one hundred years after Pope Innocent III reigned over the papacy's unchallenged supremacy. It says even more that Philip, who had been feuding with Rome over his right to tax and regulate the clergy, plotted to kidnap Boniface and bring him to Paris. Storming a palace where the pope had gone to write an order excommunicating Philip, the king's henchmen held the pope three days, until Boniface was rescued. The pope died a month later, reportedly from shock over the incident.

Another seminal moment in European history, this soap-opera affair symbolized the rising power of secularism in politics just as Abelard and Bacon signaled the start of a new secularism of the intellect. It also would trigger a century of chaos in the papacy, as it was drawn into the struggles and power plays among the emerging great powers of Europe. In 1309 a French-sponsored pope was crowned in Lyons, and established residence at Avignon in Provence, launching a 68-year absence from the holy city that nearly split the Church in two.*

Still, in the years immediately after Boniface's jubilee the overall attitude of Europeans remained positive, with trade increasing, the population growing, and frequent outbreaks of original thinking. The early 1300s was the age of Dante, who finished *The Divine Comedy* in 1321—a work filled with allusions to time, which by then was becoming a subject not just for ecclesiastics, traders, and scientists but also for poets. In his canto on paradise, Dante's narrator (also named Dante) describes the source of time—what he calls the *primum mobile,* a ring of heaven situated above all the heaven-planets. This, he says, is the invisible, unmoving force of the divine mind that directs the daily revolutions of the planets around the earth, which is of course in the center:

---

*Between 1378 and 1417, rival popes resided in Avignon and in Rome.

The nature of the universe which stills
The centre and revolves all else, from here,
As from its starting-point, all movement wills.

This heaven it is which has no other "where"
Than the Divine Mind; 'tis but in that Mind
That love, its spur, and the power it rains in here . . .

As in a plant-pot, then, time has its roots
Herein, and where the other heavens trace
Their course, thou mayst behold its shoots.

This is a poetic version of Aquinas's complementary truths: of Plato's universal, unmoving mover situated in an Aristotelian hierarchy where cause and effect are clear. The reader is invited to "behold" the inner workings of the heavens (and, in earlier cantos, hell and purgatory), and to join Dante the pilgrim in his quest to understand them as part of the great natural scheme of God's universe.

This also was the age of the poet Petrarch, the painter Giotto, and the sculptor Nicola Pisano; when dozens of universities opened; when clockmakers built the first public bell towers in major cities across Europe, and Genoan sailors set foot on the Canary Islands in the first step toward the European exploration of Africa and westward into the Atlantic Ocean.

For the calendar, however, little of interest happened during the first four decades of the 1300s. Then in 1345 the newly installed pope at Avignon, the French nobleman Clement VI (1291–1352), abruptly decided the calendar needed to be reformed.

It is not entirely clear why, though Clement seems to have been motivated by the age-old problem with Easter, and perhaps by Bacon's arguments that such an obvious error was an embarrassment to a Christendom increasingly worried over what outsiders thought about it. Whatever his reasons, this pope, known for his pomp, his extravagant living, and his patronage of the arts, dispatched letters to calendar experts on September 25, 1344, asking them to come to Avignon to consider and advise on the correction of the calendar. In his mandate the pope ordered the scholar's expenses to be paid by their local bishops.

The most important of these scholars was Jean de Meurs, an Aristotelian at the University of Paris who wrote two works in the 1320s touching on how to measure time. In one he compared the passage of time to what happens when someone plays or sings a musical piece, with its beginning and ending. Jean called this natural time, which interacts with abstract or mathematic time, which is how music is subdivided according to measures, notes, and other breaks.

In his 1345 response to Clement, titled *Epistola super reformatione antiqui kalendarii* (Letter about the reform of the ancient calendar), Jean and another time reckoner named Firmin told the pope that their ideal solution for realigning Caesar's calendar with the sun was to remove the appropriate number of days from a single year's calendar. They insisted that determining this was relatively easy, given the accuracy of the latest star charts that built on Arab and Greek texts, and on observations of Reiner, Grosseteste, and others.

Jean warned, however, that removing days from the calendar might cause turmoil for governments and commerce: quarrels about payments and contracts, and perhaps riots. He also pointed out that if the Catholics changed their calendar, they would be celebrating fixed holy dates such as Christmas on different days than Christians in the East and in other schismatic sects, setting back the still important Catholic goal of a truly universal church.

Jean and Firmin were more optimistic about reforming the 19-year lunar calendar. Recalculating Grosseteste's value for the error, they came up with a day gained every 310 years—just one and a half years off from the true Julian value of about 308.5 years. By 1345, Jean wrote, this error had accumulated to a slip of four days. They suggested that the pope restore the lunar calendar to its proper alignment by removing these days from the 19-year cycle, and order that a day be dropped thereafter every 310 years. The best year to start the reform, they said, would be 1349—the year after a leap year and the first year in the next 19-year Metonic cycle. Jean and Firmin drew up a calendar incorporating their proposed changes.

Clement VI did not formally respond to the proposal, but it seems likely that he agreed with the reforms. Indeed, this modest correction seemed well on its way to being implemented as the 1349 date approached—possibly to be followed by a reform of the solar calendar.

But it was not to be. For as 1345 passed into 1346 and 1347, the future of calendar reform—and of Europe itself—was being decided not in glittering Avignon but in a remote Genoan outpost on the Crimea, where a small fleet of trading vessels was setting sail to cross the Black Sea and then the Mediterranean. Heading off just before the winter winds at the end of the sailing season, the sailors and merchants on board these ships and others departing various Eastern ports were transporting more than spices and cloth. For in their blood a microscopic cargo was growing and spreading that would kill most of these men before they reached their destination. And with them would die what appeared to be in the time of Clement VI a nascent renaissance, one that might have solved the calendar conundrum two centuries before Gregory and Clavius.

12

From the Black Death to Copernicus

When the calendar was under discussion . . . no solution was found for the sole reason that the length of years and months and solar and lunar motion were not yet considered to be sufficiently well determined. Since that time, in fact, I have turned my attention to the observation of these phenomena.

—NICOLAUS COPERNICUS, 1543

In October 1347, two years before Clement VI's calendar reforms were to begin, the Genoan trading ships from Crimea arrived in the Sicilian harbor of Messina. Anyone watching the vessels approach would have known something was wrong. They were moving too slowly, with only a few oars beating time in the crystal-blue waters. And once the ships finally moored, an onlooker would have immediately seen the cause—that the men on board all were dead or dying. They looked like ghouls, with black boils and blotches and strange black swellings the size of apples in their armpits, necks, and groins, oozing pus and blood.

The disease was bubonic plague, and it spread like wildfire in waves north from Italy and from other coastal cities of the Mediterranean. Originating in China or India—no one knows for sure where—the bacterium *Yersinia pestis* was passed to humans by fleas carried by rats. But no one understood this at the time, which added the element of the unknown to the terror.

The plague could strike someone down in three days or less. Eyewitnesses tell of people going to bed healthy only to die before they awoke. Doctors sometimes caught the malady at a bedside and succumbed before their patient. The Florentine historian Giovanni Villani (c. 1275–1348) reportedly died in the middle of writing a sentence: *"E dure questo pistolenza fino a—"* (In the midst of this pestilence there came to an end).

In 1348 an Englishman named Henry Knighton reported to the pope that "there died in Avignon in one day one thousand three hundred and twelve persons." Others describe up to 50,000 dead in Paris and 100,000 in Florence, with daily death tolls in the hundreds in Pisa, Vienna, and elsewhere. These are almost certainly exaggerations, since few cities had populations this large to start with, and accurate estimates are hard to come by. Probably some 30 million perished in about two years—a third of all Europeans.

The horror prompted the poet Petrarch (1304–1374) to write his brother at Monrieux, after hearing that he and a dog were the sole survivors out of 35 people in the monastery where he lived.

> Alas! my beloved brother, what shall I say? How shall I begin? Whither shall I turn? On all sides is sorrow; everywhere is fear. . . . [For] without the lightnings of heaven or the fires of earth, without wars or other visible slaughter . . . well nigh the whole globe, has remained without inhabitants . . . houses were left vacant, cities deserted, the country neglected, the fields too small for the dead.★

Obviously no one in this time cared one whit about the calendar or about the planned 1349 reform. Indeed, the sudden loss of so many people plunged the continent into a deep crisis on nearly all fronts— economic, political, and intellectual—that it would not fully recover from for a century or more.

Many believed God was raining down pestilence in a latter-day Flood to punish a sinful age, including a Church that had grown too concerned with riches, fetes, and the affairs of the world. Others be-

---

★Petrarch uses as his inspiration a letter written by the Roman statesman Cicero (106–143 B.C.).

lieved these were mankind's final days and that nothing mattered, so they launched themselves into orgies and feasting. The resulting collapse of confidence in all authority eventually led to peasant revolts and riots across Europe as kings and the clergy tried, and failed, to revive the old feudal order, which was becoming moribund anyway with the rise of trade and commerce in the cities.

Europeans reeling from the plague were equally repelled by what passed for science, which had been utterly useless in stemming the disaster. When the French king Philip VI asked the medical faculty at the University of Paris for an explanation, the doctors turned not to physiology or cures but to the stars and the calendar. They actually blamed the plague on a date: March 20, 1345. On this day, they said, a triple conjunction of Saturn, Jupiter, and Mars had occurred in the 40th degree of Aquarius—which was not a good omen, apparently. These physicians also admitted to causes "hidden from even the most highly trained intellects," though it was the crossed-stars theory that became the accepted explanation for the Black Death among intellectuals. The pamphlet the Parisians produced containing this explanation was republished and translated from Latin into various vernacular languages and into Arabic, where it was "affirmed" by Arab physicians in Córdoba and Granada.

During the plague years time itself seemed to pause, as people groped to understand what had happened. This was a period when time was truly a thing to fear: the present filled with the moans and death throes of friends and family, the past haunted by those now dead—and, for those who believed the plague was punishment for mankind's sins, by past infidelities. As for the future, no one dared think about it. It was as if people were literally holding their breath, trying to keep away what one Welsh poet called the "rootless phantom," and wondering if this was truly the end of life—and therefore of time—for all humanity.

Yet even as the plague struck, Europe was reaching a critical juncture in its perception of time. Starting in the early 1300s, with the first mechanical clocks, came the conception of the hour as a secular unit of time. This was entirely separate from the old canonical "hours" used by monks, which were intended less to keep time than to demonstrate one's faith through following a highly regulated day of prayers and spiritual activities.

No one is sure when the clock was invented or by whom. In the tradition of the Middle Ages, with its de-emphasis of the individual, all we know is that sometime after the year 1300 one or several inventors fashioned out of metal several notched wheels attached to an escapement mechanism, which was then assembled with a gear train, axle, pulleys and weights, and "hands" to mark off intervals of time. The device was driven by the weights slowly dropping, which turned the notched wheel of the escapement mechanism and forced the axle to turn in regular ticks, which turned the hands. Later the weights would be replaced by coiled springs and later still by springs and pendulums.

One of the earliest drawings of a mechanical clock using weights and pulleys was by Giovanni Dondi in 1365, by which time the escapement system had been in use for decades (see page 217).

Initially the mechanical clock did little to change the medieval mind-set, which may be one reason why no one bothered to write down details about its discovery. The clock's impact was felt at first in just a few cities, and only by those people living close enough to the clock tower to read it or to hear its chimes every hour, and later every half and quarter hour or so.

Nor were these primitive clocks always reliable. They were prone to slowing down and speeding up, and to wide variations from clock to clock in what time it was and what amount of time constituted an hour. The day also started at different times in different places, depending on local custom. This meant that a traveler could hear the first hour of the day sounded on a bell in his home village at dawn, arrive at the next village in time to hear the first hour sounded at noon

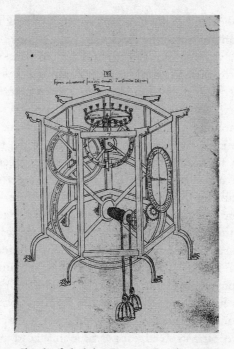

Sketch of clock by Giovanni Dondi, 1365.

and end up at his destination at midnight to hear the next first hour rung in. This added to an already confusing calendar of different names for days and different starting dates for the year.

For the new generations of clock people, however, the long-term effect was more profound than they realized, since time could now be measured objectively rather than remain subject to the interpretation of whoever was eyeballing the angle of the sun, or was deciding how long a person should work, or figuring out what time a merchant should deliver a cart full of apples to the lord's castle. This made the clock the new lord and arbiter of time for everyone within its range, whether king or priest, peasant or pope.

This new reality crept into the consciousness of various groups in different ways. For merchants and traders, clocks connected time more than ever to labor and making money—and into making the most of the present moment, since the clock starkly underlined the reality that one had only a set amount of hours and days to conduct business. In the merchant town of Siena the painter Ambrogio Lorenzetti (1290–1348) illustrated this new immediacy in a 1338 painting hanging in this trading city's town hall.* It shows Temporantia, the goddess of temperance, holding an hourglass, and sitting above people going about their business. We see the scholar in his study, the preacher in his pulpit, the advocate in his courtroom, the cobbler selling shoes, the housewife at her oven. Death is also here, underlining the need to make the most of things while there was still time.

Even the spiritually inclined embraced the clock, rejecting the centuries-long shunning among their predecessors of water clocks and other mechanical time devices. Some of them considered the clock to be a symbol of God's clocklike regulation of the universe. In 1334 the German spiritualist Heinrich Süse (c. 1295–1366) described a vision in which he had seen Christ in the form of an elaborate clock with chimes sounding out the hours. For Süse the clock mirrored the human soul, keeping its steady, inner time in accord with God's own eternal time.

Scholars too did some deep thinking about the mechanical clock. In 1377 the naturalist and prelate Nicholas Oresme (c. 1325–1382) wrote in his *Book of Heaven and the World* that the universe was like an *horloge:* a clock that was neither fast nor slow, never stopped, and worked in every season, by night as well as by day. Oresme also compared the planets and their motions to the balancing of a clock's weights by the escapement mechanism. "This is similar to when a person has made an *horloge* and sets it in motion," wrote Oresme, "and it then moves itself."

Possibly Oresme, who lived in Paris, was inspired by the large mechanical clock installed in the palace of the French king Charles V in 1362. By order of the king, after 1370 this clock became the standard

---

*Lorenzetti died of plague when the Black Death swept through Tuscany in 1348.

timepiece to set all others by—one small example from a period in which time was again being seized by the secular world. Like Julius Caesar 14 centuries earlier, Charles was assuming the role of a *magister temporis,* a master of time, using his civil authority to organize time in the most practical way he knew how, while letting it be known that as king he was arrogating to himself a power once reserved for God.

This new time consciousness was set against the backdrop of a century afflicted not only by plague and economic depression, but also by a major split in the Church—which hardly boded well for reforming the calendar.

The schism began when two popes were elected at the same time by rival groups of cardinals in 1378, one based in Rome and one in Avignon. This left the papacy in shambles and the prestige of the Church at a low point even after the papacy was restored in Rome with a single pontiff in 1417.

Meanwhile the Hundred Years War between France and England raged on, in a conflict famous for producing knights and adventurers such as the Black Prince (Edward of Woodstock, the Prince of Wales [1330–1376] and the Breton Bertrand du Guesclin (c. 1320–1380); and the occasional warrior-saint such as Joan of Arc (c. 1412–1431). This also was the age of the mercenary in Europe, when the incessant warfare of kings, despots, and popes fed a booming industry of knights, archers, and pikesmen fighting in the army of the highest bidder. It was not unheard of for mercenaries to switch sides in the middle of a battle or campaign if the "enemy" offered more gold. And when a campaign ended, the mercenaries would often terrorize the peaceful countryside.

Predictably, learning and scholarship were not high priorities for warring kings and prelates. Nor did scholars produce much original work during a period where the great surge of intellectualism after the founding of the universities and the arrival of new knowledge was wearing thin, with university curricula and approaches to learning becoming in many cases stale and in need of reinvigoration.

During this troubled time, the Church attempted to repeat the success at Nicaea so many centuries earlier by calling a series of great councils. The first of these, held from 1408 to 1418 in Constance—on the border between Germany and Austria—attempted to deal with the schism, finally finding success when a single pope was elected in 1417 to take up residence in Rome. At the same time, at least one important figure at the Council of Constance tried to interest one of the "antipopes," John XXIII, in reforming the calendar. This was Cardinal Pierre d'Ailly (c. 1350–1420), an astronomer and former chancellor of the University of Paris who presented a treatise at Constance detailing the usual laments about faulty measurements and Easter. In his *Exhortatio super correctione calendarii*—"plea to correct the calendar"—he offered his reform ideas, which were mostly a rehashing of Grosseteste, Sacrobosco, and especially Roger Bacon, who by now was an acknowledged master on this subject, over a century after his death in obscurity.

Pope John responded by issuing a decree in 1412 to correct the drift in the lunar calendar by removing four days, the solution suggested by Jean de Meurs in 1345. But amidst the turmoil of dueling popes, John's edict was ignored. The proposal also foundered because astronomy still lacked precise planetary and star charts with which to calculate a proper correction. Cardinal d'Ailly himself admitted "that the true length of the year is still not known to us with complete certainty." Other efforts at reform failed to catch on at Constance in 1415 and 1417.

In 1436 the astronomer and philosopher Nicholas of Cusa (1401–1464) delivered to another council—this one held at Basel, in Switzerland—yet another compendium of the problem in his *De correctione Kalendari*.* Working with a commission of experts on calendar reform,

---

*Nicholas of Cusa was also the author of two of the earliest scale maps of land areas,

he proposed canceling seven days in 1439 and thereafter adding one day every 304 years. But critics objected again on the grounds that the astronomy remained too uncertain. They also worried that removing days would create economic confusion with deadlines, contracts, and interest payments thrown into disarray. Anyway the Church remained far too distracted with its own affairs to affect a change.

By the mid-1400s Europe was beginning to recover from the disastrous effects of the plague. Politics remained unsettled, with more campaigns and skirmishes. Byzantium fell in 1453, when the Turks breached the once invincible walls of Constantine's ancient city using a newly arrived invention, the cannon.

As the fifteenth century waned the Church finally began to put its house in order, turning to serious debates over its role in a Europe that was quickly becoming more secular, and over an outdated medieval dogma being challenged by the new philosophy of the nascent Renaissance. Humanism, a movement that emphasized human welfare, values, and dignity stood in opposition to the medieval emphasis on spiritualism, pageantry, and the absolutism of the Church and papacy. At the same time, the awe felt by earlier generations of intellectuals for texts by Ptolemy, Aristotle, Euclid, and other ancient masters, who previously had been the last word on science and philosophy, were beginning to give way to a new curiosity to use this knowledge as a basis to explore the world and to test the old ideas.

This was mirrored across the social spectrum as Europe's economy revived between 1460 and 1500, and the Europeans began to expand their thinking and influence in the world. In commerce, European

---

including attempts at the longitude and latitude of Europe. And he repeated the idea that the earth rotates, as suggested by Aristarchus, Aryabhata, and others.

ships sought markets farther afield than ever, with explorers heading off in all directions. In 1470 Portuguese sailors discovered the Gold Coast while tracing the coast of Africa in newly invented caravels. In 1486 they found Angola. Six years later Ferdinand and Isabella of Spain bankrolled Christopher Columbus to trek across the Ocean Sea. That same year the Spanish defeated the last of the Arabs occupying Iberia.

In warfare, engineers invented or borrowed and improved upon new methods for making armor and battlements. They learned to use cannon and gunpowder, borrowed from the Turks and Arabs, who had borrowed them from the Chinese. At the same time tinkerers, scientists, and entrepreneurs became both more common and more bold, in the spirit of Roger Bacon. The resulting inventions included the printing press, in about 1470, perhaps the most important creation of this era.

Among other things, the printing press allowed calendars to be mass-produced, bringing for the first time a standardized, easy-to-read rendering of days, weeks, months, and holidays to people other than astronomers, ecclesiastics, kings, and tax collectors. The earliest printed calendars used symbols so that illiterates could count the days, and employed illustrations of saints and pictures to represent feast days. In the calendar, a "Farmer's Calendar" printed in Zurich for the leap year 1544, the days are depicted with black triangles, except for Sunday, which is red. Other symbols list the passing progression of the zodiac; kalends, nones, and ides; phases of the moon; and saint's days. Easter is marked on April 13 with a cross.

Toward the end of the century, Leonardo da Vinci and other Renaissance inventors were at work. So were a new generation of artists— Leonardo once again, Michelangelo, and Raphael, to name only three—who applied the science of the ancients to a new visual sense of perspective, beauty, and symmetry in composition that blended reality and a classic Greek sensibility for perfection and beauty in paintings and sculptures.

Meanwhile the calendar at the turn of the sixteenth century had drifted away from the true seasons of the earth by over twelve days since Caesar, and over nine days since the Council of Nicaea. In 1500 no one could measure this error exactly, but every intellectual acquainted with mathematics, astronomy, or theology knew about it. But how to fix it? And who decided?

These questions popped up at yet another Church-wide council begun at the height of the Renaissance in Italy. In 1512, Pope Julius II convened the Fifth Lateran Council (1512–1517) at the Lateran Palace in Rome, presided over by Julius and his successor, Leo X. Again, calendar reform was not a major agenda item in a meeting called to settle issues such as how much power the pope wielded over kings, and the raising of a Christian army to combat the Turks—whose troops after taking Byzantium had surged into Europe to seize Greece and much of the Balkans.

Calls for calendar reform had been increasing, however, even as the difficulties of how to accomplish this became more complicated. For instance, should the proper date for the equinox be based on the year of Caesar's reform, the time of Christ, the Council of Nicaea, or the creation of the world? What was the correct meridian on which to base the Easter calculation: Rome? Jerusalem? And what happens when the equinox falls at the end of the day in Rome and lands on the next day in the Holy Land?

A number of astronomers tried to deal with these questions by improving the charts measuring the equinoxes to make them more accurate. None got it right, though. Indeed, as the new charts circulated, the glaring errors in the calendar became more widely known, and a persistent source of embarrassment for the Church. The wide dissemination of printed calendars, such as the oft-copied "Shepherd's Calendar," first issued in 1493, added to a sense of urgency as more people than ever used the Church's calendar for business, governing, and personal planning.

In 1514 Pope Leo X invited the period's greatest expert on the calendar, the Dutch astronomer, physician, and bishop Paul of Middelburg (c. 1450–1533), to head up a commission on reforming the calendar. A few years earlier, in 1497, Paul had written a strident tract to the pope demanding he reform the calendar. In 1513 he wrote another impassioned tract opening with letters appealing to Leo, the Holy Roman Emperor Maximillian I, the College of Cardinals, and the Lateran Council.

As head of Leo's new commission, Paul started by criticizing past reformers, particularly those who wanted to drop days from the year to correct the calendar's drift. He proposed fixing the calendar not by dropping days, but by changing the date of the vernal equinox to March 10—which he wrongly estimated to be the proper date for his time. He suggested that in the future the equinox be allowed to drift through the calendar every 134 years—this (wrong) number coming from a set of astronomic charts considered highly accurate at the time: the Alfonsine Charts, completed in 1272 by astronomers at the Castilian court of King Alfonso X (1221–1284). Paul also proposed a slight rearrangement in the lunar calendar, including dropping a day every 304 years—and the naming of lunar months after the ancient Egyptian months, to avoid using Moslem or Jewish names. The proposals were to be considered in December 1514, with the changes to be made retroactive to January 1, 1500, when the astrologically minded Paul noted a mean conjunction of the sun and moon had occurred along the Rome meridian at noon on the first day of this important jubilee year. Surely, said Paul, this was a sign from God concerning his desire to reform the calendar.

Leo X ordered letters dispatched in 1514 from the papal curia to all important Christian monarchs asking for opinions on the proposal from their astronomers and other experts. But only a few responded, in part because they were not given much time before the decisive meeting that December.

The British for one did not respond, though four letters from Leo X to Henry VIII survive in the British archives, all apparently unanswered. On July 21, 1514, Leo's first letter describes the problem and laments that "Jews and heretics" were laughing at the flawed Christian calendar. Leo asked Henry to send his best astronomer or theologian to Rome, or else a written version of their views on the calendar. Two years later, on June 1, 1516, the pope's second letter complains of the poor response to the first missive, which led to the cancellation of the planned December calendar conference. He asks Henry to respond in time for the next session of the council, scheduled for later that year. Two other letters that year repeat the pope's request, which presumably went out also to other kings who failed to answer. This lack of interest apparently doomed Paul's effort at reform.

One papal letter that was not ignored prompted a response from a young German-Polish astronomer then living in Frauenburg on the Baltic coast of Poland—listed as a respondent by Paul under the name Nicolaus Copernicus Warmiensis. Known as Mikolaj Kopernik in Poland, we know him by his Latinized pen name of Nicolaus Copernicus (1473–1543).

In his early forties when the papal letter arrived, Copernicus was a canon at Frauenburg's cathedral in this often cold, stormy coastal town near the Gulf of Danzig in what was once East Prussia. A man with a long nose, wide eyes topped by arching eyebrows, and a quiet demeanor—at least this is how he looks in his self-portrait—Copernicus had settled here after years of studying and teaching at universities in Krakow, Bologna, and Padua, where he had earned degrees in law and medicine. In 1500 he had traveled to Rome for the jubilee. He also met and worked in these early years with a number of leading scholars, with whom he kept in contact for the rest of his life.

Around 1506, when Copernicus returned to the Frauenburg area, he began the astronomic studies and observations that would occupy the rest of his life. By 1512 he had written a short, unpublished manuscript outlining his early thinking about his planetary theories.★ Two years later, in 1514, the pope's missive arrived, an event alluded to by Copernicus himself in his 1543 dedication for *De revolutionibus,* in which he also tells us his response to the pope's inquiry:

> For not many years ago under Leo X when the Lateran council was considering the question of reforming the Ecclesiastical Calendar, no decision was reached, for the sole reason that the magnitude of the year and the months and the movements of the sun and moon had not yet been measured with sufficient accuracy. From that point on I gave attention to making more exact observations of these things and was encouraged to do so by that most distinguished man, Paul [of Middelburg], Lord Bishop of Fossombrone, who had been present at those deliberations.

After Paul's commission sputtered out, the matter of the calendar was dropped again for over 60 years during yet another tremendous upheaval in the Church: the rise of Protestantism.

It was born during the final year of the Lateran Council, in 1517, when Martin Luther (1483–1546) tacked a document on the door of the cathedral at Wittenberg in Germany, complaining about the sale of indulgences by the Church. Luther at first did not intend to start a revolution, though he followed his act of defiance by preaching what amounted to a direct challenge against Rome. Insisting that the Bible should be the sole authority in the Church, and that salvation lay solely in faith—the first denying the pope's authority and the second contradicting core Catholic doctrine—Luther touched a powerful nerve of discontent. In the 1520s he broke off with Rome to head up a movement that swept through Europe, attracting as many as half of all Christians in the West by midcentury.

This in turn incited a backlash of conservatism in the Catholic

---

★This was finally published in 1530.

Church, and an intense counterreformative effort by the papacy and loyal Catholic monarchs to stamp out Protestantism. It included a new Inquisition launched by Pope Paul III in 1542 and the founding of the Jesuits in 1540, in part to create religious and theological stalwarts to argue against and fight the spread of Protestantism.

During these years of upheaval Copernicus worked quietly in Frauenburg: writing, taking astronomical observations, fulfilling his duties as a cathedral canon, and tending to the occasional patient as a medical doctor of some renown.

Apparently he lived in rooms occupying a three-story turret set in the cathedral's thick surrounding walls, built in the fourteenth century as a defense against the pagan-leaning Slavs. Standing high above a small freshwater lagoon just off the Gulf of Danzig, the turret gave the canon an excellent view of the shoreline, the deep-blue Baltic, and the stars. He used relatively simple astronomical instruments—an astrolabe, an armillary sphere,* and various devices to measure the altitudes of celestial objects, including the sun. Copernicus later published some of these observations in *De revolutionibus*. He also jotted them over the years of quiet study in his tower rooms into the flyleaves and margins of books in his library. It was in these rooms that Copernicus worked and reworked the opus that became *De revolutionibus*—which included attempts to properly measure and calculate the length of the year.

Copernicus tried to fulfill his promise to Leo X by making his own fresh calculations based in part on his own sightings, and by using those made by Greek and Arab astronomers over the centuries. Summing up his findings and thoughts in *De revolutionibus,* he begins a section called "On the Magnitude and Difference of the Solar Year" by first explaining the difference between the two types of "years" measured by astronomers.

First is the seasonal or tropical year, which is the time it takes for

---

*This was a concentric series of metal rings, each representing a planet's orbit. Arranged in a sphere, they could be used to measure and calculate planetary movements.

the seasons to cycle through and start again. This has been the "year" we have referred to throughout this book and which is the basis for our season-based calendar year. It is determined by measuring the length of time between vernal equinoxes, when the planes of the equator and the sun's ecliptic intersect in the spring. The other year is the "star" year, also called the sidereal year, which measures the time it takes for the earth to revolve around the sun back to an exact starting point in space. The difference in these two "years," we now know, is about twenty minutes, with the tropical year running faster each year than the sidereal year. Known as the precession of the equinoxes, the phenomenon of a slower tropical year was first discovered by Hipparchus in ancient Alexandria, though it took until Newton for astronomers to understand its cause: gravitational pulls and tugs from the sun and moon, against an earth that is not a perfect sphere—which cause the earth's axis to wobble slightly.

But Copernicus did not know this. Nor did Ptolemy in A.D. 139 or the Arab astronomer al-Battani in 882, whose calculations Copernicus trusted and used to compare his own observations for the tropical year:

> We too made observations of the autumn equinox at Frauenburg in the year of Our Lord 1515 on the 18th day before the Kalends of October. . . . The time between our equinox and that of al-Battani there were 633 Egyptian years and 153 days and 6¾ hours. . . . But between the observation made by Ptolemy at Alexandria, there were 1376 Egyptian years 332 days ½ hour. . . . Therefore during the 633 years between al-Battani and us there have fallen out 4 days 22¾ hours, or 1 day per 128 years; but during the 1376 years after Ptolemy approximately 12 days, i.e., 1 day per 115 years.

Naturally Copernicus was perplexed by the difference between Ptolemy's and al-Battani's numbers, not realizing that both of their measurements were wrong. This led to an erroneous conclusion blaming the discrepancies on irregular motions of the earth that he believed affected the tropical year as measured by the equinoxes.

Still, Copernicus came up with a remarkably accurate measurement of the tropical year: 365.2425 days, or 365 days, 5 hours, 49 minutes, and 29 seconds: one of the closest estimates yet to the true value (at that time) of about 365.2422 days—365 days, 5 hours, 48 minutes, and 46 seconds. He also provided measurements and data that would become important four decades after the publication of his tome as Pope Gregory's calendar commission struggled to come up with an acceptable measurement of the year.

Given the confusion over the supposed "irregularity" in the tropical year, Copernicus preferred to use the sidereal measurement, which he estimated to be 365 days, 6 hours, 9 minutes, and 40 seconds, or 365.25671 days. This is about 30 seconds greater than the true value. "But also in the case of the astral or sidereal year an error can come about," he admits, "but nevertheless a very slight one and far less than the one which we have already described" for the tropical year.

Copernicus labored over his opus for over 30 years but remained reluctant to publish *De revolutionibus,* knowing his sun-centered hypothesis would not be well received by traditionalists both in the Church and in academia. Indeed, for millennia humankind had assumed the earth was the center of the universe—a theory "proved" by Ptolemy and every other major astronomer, ancient and modern. To say otherwise was laughable to people of that day, even if it came from a man such as Copernicus, who was widely revered as an expert on astronomy. It took considerable persuasion by Copernicus's friends and admirers—led by his disciple and colleague Georg Joachim Rhäticus (1514–1576)—to talk the elderly Copernicus into finally publishing *De revolutionibus*.

He did so shortly before he died at age 70, but not before Copernicus added a dedication to Pope Paul III, acknowledging that his views were controversial but begging the indulgence of the Church to consider the science behind his hypothesis.

According to a friend who stood by his deathbed, the old astronomer finally got to glimpse his published masterpiece on the very day he succumbed to a months-long illness, on May 24, 1543. "He had lost his memory and mental vigor many days before," wrote this friend

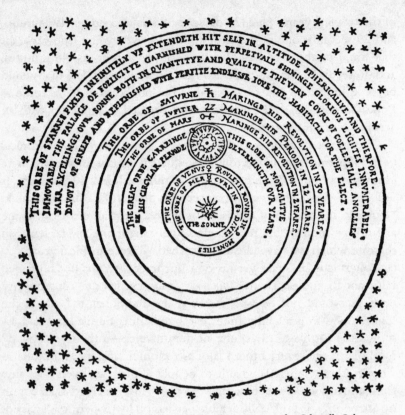

The Copernican Solar System, *Perfit Description of the Celestiall Orbes,*
Thomas Digges, 1576.

in a letter to Rhäticus, "and he saw his completed work only at his
last breath upon the day that he died."

Despite Copernicus's fears, his book initially attracted little controversy. Very few people could understand it, and those who did went

along with a preface added to the book without Copernicus's permission that described its contents as mere conjecture rather than probable fact. An exception was the vehement reaction of Luther and the Protestants. As biblical purists, they viewed any deviation from the scriptures as subversive, and refuted Copernicus's sun-centered planetary system with passages in the Bible that seemed to imply that the earth stands still while the sun moves. "The fool wants to overturn the whole science of astronomy," said Luther, "but, according to the Scripture, Joshua bade the sun and not the earth to stand still."*

For seventy years the Roman Church remained silent about Copernicus. Then Galileo Galilei (1564–1642) began peering in the early 1600s through his newfangled telescope at the planets and stars, leading him to publicly endorse Copernicus's sun-centered hypothesis in 1613—a contention that two years later led to Galileo being denounced to the Inquisition as a heretic. He cleared himself of the charge, but created such a sensation that the Church officially investigated the Copernican theory early in 1616, with Church authorities ordered to examine the fundamental Copernican assertion "that the earth is the center of the universe and is wholly stationary," and "that the sun is not the center of the universe, and is not stationary, but moves bodily and also with a diurnal motion." On February 24, 1616, the Qualifiers of the Holy Office concluded that a sun-centered theory was "foolish and absurd in philosophy, and formally heretical, inasmuch as it expressly contradicts the teachings of many passages of Holy Scriptures."

This came at a time when the Counter-Reformation had sharply focused the Catholics toward following strict dogma. This rigidity led the Church to make a profound error when the Inquisition in 1635 forced Galileo to abjure the heliocentric theory or face torture or possible execution. As it turned out, this was one of the last great attempts by the old order of the Middle Ages to subjugate science to dogma, and the sacred to the profane.

---

*Luther is referring to an Old Testament story in which the Jewish prophet Joshua during a battle commanded the sun to stay in the sky.

But this came later. In the years immediately following the publication of *De revolutionibus* astronomers reading it were less interested in the sun-versus-earth debate than in studying and using Copernicus's observations and general theories on planetary motions—including his estimates of the length of the year and his measurements of lunar phases. Indeed, the work of Copernicus, combined with other astronomic charting of the era, set the stage for two virtually forgotten men—a mathematician from Bavaria and a physician from southern Italy—and a pope named Gregory, who would finally come up with a most elegant solution to fix the calendar, and even more importantly, to enact it.

13

Solving the Riddle of Time

The patriarch has also subscribed to our calendar and admitted that it is very good. I hope that it will soon be published, because the Pope is quite eager.

—CHRISTOPHER CLAVIUS, 1581

None of the three men responsible for fixing the calendar was a conqueror, notorious lover, heretic, or lone monk pondering the cosmos from a cell in a monastery. They were not even particularly flamboyant, and certainly not freethinkers in the spirit of a Bacon or even a Paul of Middelburg—all of which might account for their success.

They included an obscure physician from the toe of Italy who was the genius behind the reform, a Jesuit astronomer famous for being wrong about many of his most cherished theories, and a lawyer turned pope remembered as much for his failures as for his successes. Each contributed to the reform named for one of them, and each in the story of his role offers an explanation for why the calendar was finally fixed 1,627 years after Caesar launched it, and after so many centuries of false tries and frustrations.

The doctor was Aloysius Lilius.* Born about 1510 to a family of modest means, little is known about Lilius—the *"primus auctor"* of the Gregorian reform, according to a prominent member of the calendar

---

*Luigi Lilio in Italian.

233

commission. He is said to have studied medicine and astronomy at Naples, settled in Verona, and taught at the University of Perugia before returning late in life to his hometown of Ciro, in southeastern Italy, where he concocted the solution to the calendar conundrum and designed the reforms. Indeed, if the pope had offered a prize for solving this age-old problem—as the British later offered a prize of £20,000 to anyone who solved the ancient puzzle of determining longitude at sea—Aloysius Lilius could have rightly claimed it. But this forgotten man never had the chance. For before his solution could be presented in 1576 to the pope's commission in Rome, Lilius took ill and died.★

After Lilius's death, his brother Antonio, also a physician acquainted with astronomy, presented Aloysius's plan to the calendar commission. They quickly embraced it as their leading proposal, admiring it for its simplicity, elegance, and avoidance of controversy. Antonio stayed on in Rome as his brother's representative. Later he was the recipient of what passed for a discoverer's "prize" in the sixteenth century: a 1583 bull from Pope Gregory that granted him the exclusive right to publish the reformed calendar and its new rules for a period of ten years. This potentially lucrative license was later rescinded when Antonio failed to produce enough copies fast enough to meet the demand, a delay that nearly derailed the reform.

The second prime mover was the Jesuit astronomer Christopher Clavius (1537–1612), the man behind the scenes who championed Lilius's ideas (after an initial skepticism) and shepherded the reform through the minefields of scientific and ecclesiastic controversy before and after 1582. Until he died in 1612, Clavius worked hard to defend and

---

★Some accounts say Lilius died in Rome.

explain the new calendar, ensuring that it would spread beyond the handful of countries that initially accepted it.

As a prominent public figure in Rome during the late sixteenth and early seventeenth centuries, more is known about Christopher Clavius than about Lilius. Yet little exists to flesh out who he really was. In a portrait of Clavius rendered in 1606 he is dressed in a simple Jesuit robe and a four-cornered hat. A portly, satisfied-looking man with a pudgy, bearded face, he looks sympathetic, even kind—the sort of scholar who is serious but never stuffy, smart but not precocious; one that students are fond of, and one that politicians and prelates feel comfortable assigning to commissions.

To his contemporaries Clavius was a revered sage of math and astronomy, acclaimed as "the Euclid of his times" in part because he penned a widely used translation of the original Euclid, along with several other works considered important in his day. Even the era's greatest scientific firebrand, Galileo Galilei, came to him for validation of his telescopic observations of the moon, sun, and planets. Clavius hailed them as important to astronomy, but since he was a confirmed defender of Ptolemy he disagreed with Galileo's interpretation that craters on the moon, Venus passing through its phases, and moons around Jupiter suggested Copernicus was correct. Clavius also has the distinction of having his face inscribed on a marble relief on the base of Gregory XIII's imposing statue in St. Peter's (probably Clavius) which shows a priest handing the pope a copy of the calendar reform.

Yet Clavius today is nearly as obscure as Aloysius Lilius. In part this comes from the bad luck to have lived between Copernicus—Clavius was five years old when *De revolutionibus* was published—and the young Galileo, who burst onto the scene in Clavius's final years. But more than anything, Christopher Clavius is obscure because he adhered to a worldview that turned out to be wrong. This made him a hero to traditionalists while he was alive, but a fool to those who came later.

Clavius was surprisingly young when Pope Gregory named him to his new calendar commission, convened in the mid-1570s. Born on

Christopher Clavius about 1606

March 25, 1537, in the Bavarian town of Bamberg, Clavius's life to us is a blank page until he joined the recently formed Society of Jesus—the Jesuits—in Rome on April 12, 1555. Studying in Rome and then at the University of Coimbra in Portugal, Clavius returned to Rome in the early 1560s to finish his education and then to teach at the Jesuits' own Collegio Romano, where he became a professor of mathematics. But for a few short trips, he would remain in Rome until his death.

As a mathematician and astronomer, Clavius was a minor figure, notable mostly for his work on Euclid, algebraic notation, and the calendar—and for his staunch defense of an earth-centered universe. Yet Clavius was flexible enough to constantly update his own theories to incorporate Copernican data and Galileo's observations, attempting to squeeze it into an increasingly strained Ptolemaic interpretation.

Clavius's willingness after 1582 to at least consider new ideas as Rome's senior astronomer seems to have exercised a restraining influence on the inevitable showdown between the ideas of Copernicus and those of Ptolemy, primarily benefiting the young Galileo, whose reputation was enhanced by Clavius's support of his telescopic discoveries. Galileo judged Clavius to be "worthy of immortal fame," and forgave him for rejecting the Copernican theory, a shortcoming he blamed on the old man's age.

Others were not so forgiving. In 1611 the English poet and satirist John Donne (1572–1631), a former Catholic in this sometimes virulently anti-Catholic kingdom, penned a vicious satire of the Jesuits and their founder, Ignatius Loyola (1491–1556), titled *Ignatius His Conclave*. Donne describes Loyola in hell trying to convince Satan to reject Copernicus because the Polish astronomer had not done enough to obfuscate the minds of men and therefore keep them from the truth. In the midst of this the poet mentions Clavius, whom he could not place in hell because in 1611 the old astronomer was still alive. But Donne did have his Loyola tell the dead Copernicus about a candidate possibly more "worthy" for the netherworld, describing among other things Clavius's work on calendar reform, which the English, as Protestants, considered tainted because it came from Rome:

> If therefore any man have honour or title to this place in this matter, it belongs wholly to our *Clavius*★ who opposed himselfe opportunely against you, and the truth, which at that time was creeping into every man's minde. Hee only can be called the Author of all contentions, and schoole-combats in this cause; and no greater profit can bee hoped for heerein, but that for such brabbles, more necessarie matters bee neglected. And yet not onely for this is our *Clavius* to be honoured, but for the great paines also which hee tooke in the *Gregorian Calendar,* by which both the peace of the Church, and Civill businesses have beene egregiously troubled: nor hath heaven it selfe escaped his violence, but hath ever since obeied his apointments: so that *S. Stephen, John Baptist,* & all the rest, which have bin commanded to worke miracles at certain appointed dates . . . do not now attend till the day come, as they are accustomed, but are awaked ten daies sooner, and constrained by him to come downe from heaven to do that businesse.

The final person in our troika was born Ugo Buoncompagni (1502–1585). The son of a noble family in Rome, he became a prominent ecclesiastic lawyer and senior papal official before being elected Pope Gregory XIII at age 70, on May 14, 1572. One of several pontiffs in the sixteenth century who worked to rebuild the authority of the Church and to reform its worst excesses, he was zealous in trying to stamp out Protestantism, chiefly by lavishing money on building up Catholic colleges across Europe, and by launching Church reforms in Germany, Poland, and Belgium. He also dispatched Jesuit missionaries to countries such as India, the Philippines, and China, where European ships had begun to sail with some regularity.

But Gregory also suppressed knowledge that failed to agree with Church dogma, establishing an infamous index of banned books that

---

★This is a satirical use of the name given to Clavius by the Jesuits, who called him "our Clavius."

later listed Copernicus's *De revolutionibus*. He also supported military efforts by Catholic monarchs against Protestants, and connived in attempts to undermine England and Queen Elizabeth I—including ill-conceived military ventures to thwart English efforts to conquer and dominate Ireland. But all this pales against Gregory's infamous response to the slaughter of thousands of Huguenots in Paris that began on St. Bartholomew's Day, 1572. Hearing the news, the newly installed pope is said to have ordered a *Te Deum*—a hymn of praise to God—and issued a medal.*

In Rome Gregory supported grandiose building projects; he also was known as a man who enjoyed pomp and celebration, nearly bankrupting the Vatican treasury with his edifices and fetes. His tenure as ruler of the papal state—a swath of land running across the middle of Italy and governed directly by the Vatican—was marked by peasant riots over steep taxes and by a rise in banditry and lawlessness, which he proved incapable of stopping.

But most of this has been forgotten, with Gregory chiefly remembered as the pope who finally corrected time, a feat that begs the question: why *this* pope?

Probably his motivation came from the same zeal he devoted to promoting education and putting the Church back onto a more sound intellectual track. But it also came from the lawyer Ugo Buoncompagni's systematic attempts as pontiff to enact reforms approved by the various church councils, particularly those passed at the various sessions of the Council of Trent (1545–1563), where Buoncompagni served as Pius IV's deputy and may have drafted some of the decrees. One of these ordered the reissuance of the mass book and breviary—the Catholic list of daily hymns and ceremonies—which implied the need for an updated calendar. Indeed, the first words in the momentous 1582 bull announcing the calendar reform do not claim the au-

---

*Catholic apologists insisted he did so without knowing the extent of the massacre, and that he actually wept when he heard the truth.

thority of science, the Church, or even God, but the decree of Trent, as if this legalistic sanction mattered most to this old lawyer-pope:

> Among the most serious tasks,★ last perhaps but not least of those which in
> our pastoral duty we must attend to, is to complete with the help of God
> what the Council of Trent has reserved to the Apostolic see.

As the pace of reform quickened, the story of the calendar returns to the same city where Julius Caesar had launched his calendar 16 centuries earlier—though it could hardly have been more different.

Rome in the sixteenth century had ceased long before to be important as a commercial, political, or intellectual center. Nor did the Roman Church wield the all-embracing authority it once had enjoyed as Europe's religious overlord, now that Protestantism had broken up its monopoly of the spirit, and kings and princes had eclipsed its influence in the realms of politics and finance. Still, the Church remained the only force in Western Europe capable of exerting anything like a universal authority. It also had been the guardian of the calendar for centuries, for better or for worse, and was now riding a certain momentum from years of reform talk and council decrees aimed at making a fix.

Rome itself in the 1570s looked ruined and exhausted, its ancient monuments, palaces, and temples shattered and half buried by dirt and rubbish, its ancient walls and columns picked apart for centuries and incorporated into a disconcerting hodgepodge of old and new. Even the once mighty Forum, where 16 centuries earlier Caesar had stood up to announce that he was establishing a new calendar, was now called the Campo Vaccino, the "Cow Pasture." Buried under eons

---

★Like all bulls, this one was named after its opening lines. In Latin, this has been shortened to *Inter gravissimas*.

of trash and dust, and mostly dismantled for its marble and bricks, this place that had been the center of the Roman world was now the domain of bovines chewing tufts of grass growing around broken columns and archways.

The Eternal City that Clavius and Gregory lived in during the years of the calendar commission stood inside the sprawling ancient walls built in the third century by Emperor Aurelian. Diminished now from as many as a million people in imperial times to perhaps sixty thousand—though in the 1570s it was beginning to grow again—the city's inhabited areas were clustered near the Tiber, where those who stayed through the barbarian invasions had moved for easy access to water after the aqueducts were cut. This left large sections inside the walls empty of people. These vast stretches of space were used for vineyards, gardens, garbage dumps, and pastureland, and were marked here and there by scattered farmhouses and convents. Forests grew on the slopes of the Palatine, Caelian, and Aventine Hills. Deer and boar ran wild amidst the ruins of ancient villas covered with ivy and trees in which hundreds of pigeons squawked and fluttered.

Because of the water problem and the location of St. Peter's near the river, Rome's center had shifted north from the Forum to the C-shaped bend in the river between the Capitoline Hill to the south (just above the Forum-turned-pasture) and the Piazza del Popolo to the northwest. Still very much a medieval city, Rome in those days was a confusing knot of narrow, winding, fetid streets filled with people, animals, dung, dust, and sewage and edged by brick houses, shops, stalls, and offices. This was broken up here and there by piazzas and by a scattering of new Renaissance churches, including St. Peter's Basilica with its half-finished dome by Bramante and Michelangelo. Rome's fractious noble families had recently erected a number of splendid new palaces and villas, many of them on hills with breathtaking views of the city.

Another new building project was an extensive upgrading of Christopher Clavius's own Collegio Romano, which Gregory XIII took on as part of his efforts to improve Catholic universities. He lavished funds and support on the previously struggling Collegio, in part because of his close ties to his favorite astronomer, who made a special pitch for improvements in the departments of mathematics and astronomy.

The pope's attention to Roman education was long overdue. Before his improvements, the Collegio Romano had been one of two clearly second-rate outposts of learning in a city known for raucous local politics, pilgrimages, indulgences, and papal fetes, but not for intellectual pursuits. Rome in the 1570s still lacked any meaningful tradition of universities and scholarship. Nor did its officials offer much public support for scientific or technical research—unlike cities such as Florence, where the ruling Medicis hired Galileo as their court mathematician in 1610, or the Holy Roman Imperial court, which commissioned the astronomers Tycho Brahe (1546–1601) and later Johannes Kepler (1571–1630) to advise Emperor Rudolf II of Bohemia.

In the lovely Tuscan village of Siena is a painting of Pope Gregory XIII crowned and enthroned, leaning forward and listening intently to a scholar on the calendar commission describe the error in Caesar's calendar. This man looks like Clavius as depicted when he was older, with a white beard and four-cornered hat. Pointing to a picture of the zodiac on the wall he is explaining to the pope the difference between the Julian calendar, marked on a band outside the zodiac, and the true seasonal year, portrayed on the inside. He stands amidst members of the commission, some of whom are dressed in the flowing robes, broad-brimmed hats, and priestly hoods popular at that time in Italy. Seated around a table, the commission is surrounded by books and astronomic tools, including an armillary sphere the scholarly speaker is manipulating with his left hand as he points to the zodiac chart with his right.

The names of the members of the commission that worked through

The Reform of the Calendar
Pope Gregory XIII Meets with His Calendar Commission, c. 1581

the 1570s and early 1580s were not recorded, except in the final report presented in 1581 to the pope—which is probably the meeting depicted in the Siena painting. Nine individuals signed this report, presumably all of them members of the commission, though one seems to have been simply a witness. The signatories included a cardinal, a bishop, a former Syrian patriarch, a man from Malta, a French lawyer, a Spanish historian and theologian, a physician, and two scholar-scientists.

The cardinal and the bishop were senior church officials now all but forgotten. They are Cardinal Guglielmo Sirleto (1514–1585), a scholar, Hellenist, and contender for the papacy, who served as president of the commission, and Bishop Vincenzodi Lauri of Mondovì. Why they were chosen is unknown, though in Sirleto's case the appointment of someone so senior and respected was clearly a signal by the pope to the Vatican bureaucracy and to everyone else within the Church that Gregory was serious about reform. Sirleto and Lauri may also have been experts on the Church calendar and its history, and on the deliberations of Church councils.

The patriarch was Ignatius of Antioch, a Jacobite Christian from Syria who had arrived in Rome in 1577 or 1578 to seek a personal reconciliation with the Roman Church. A refugee from the still-mysterious East whom some suspected was a fake—until he was confirmed as genuine—Ignatius was knowledgeable in mathematics and medicine, and he brought to the commission an Eastern perspective on astronomy and the calendar. He provided Clavius and the scientists with useful comments on their proposed reforms, written in Arabic and translated into Latin. He signed the 1581 report in Arabic and Syriac.

The man from Malta, Leonardo Abel, seems to have signed the final report just to serve as a witness to Ignatius's signatures, apparently because he was fluent in Arabic. The French lawyer signed his Latinized name as Seraphinus Olivarius Rotae, an *auditor Gallus* who may have been summoned to help the commission sort through the many legal implications of the reform for both canon and civil law. The Spaniard was Pedro Chacón, who probably advised the committee on past and present papal and Church pronouncements on the calendar, and on the critical issues of Easter and saint's days. He also authored some of the key documents of the commission.

The scholar-scientists included the Dominican friar Ignazio Danti (1536–1586), the second most famous commission member after Clavius. A mathematician, astronomer, cartographer, and artist, Danti was a professor of mathematics at Pisa and later at Bologna. Summoned to Florence he also worked on astronomic projects under Grand Duke Cosimo I (Cosmos de' Medici), preparing maps, an enormous terrestrial globe, and instruments he used to observe the vernal equinoxes in 1574 and 1575. From this he came up with the length of the year as 365 days, 5 hours, and 48 minutes. Comparing this to Ptolemy's erroneous calculation of 365 days, 5 hours, and 55 minutes,

Danti joined Copernicus and other astronomers by concluding that the tropical year was variable. After a falling-out with Cosimo's son, Danti relocated to Bologna, where he measured the solstices in 1576 with a gnomon he built in the church of St. Petronius. He used this data to confirm the error in the Julian calendar and its drift against the true year.

In 1580 Danti was summoned to Rome by the pope to join the commission, and also to design the frescoes and astronomic instruments in a new building devoted to astronomy and to calendar reckoning. Known as the Tower of the Winds, this 240-foot tower north of St. Peter's dome and above the Vatican archives, was built between 1578 and 1580 and decorated with Danti's designs between 1580 and 1582. These included a series of enormous frescoes of the four winds, rendered in the style of Titian as voluptuous cupids flanked by images of astronomers at work. Danti also equipped the main room of the tower with an enormous anemometer (wind gauge) attached to a weathervane. He etched into the floor a map of the stars and zodiac, situated so that a small hole in the wall would shine a ray of sunlight onto the map, varying according to the seasonal angle of the sun. This created in the Tower of the Winds a crude seasonal calendar. In 1583, after the reform, Danti was named bishop of Alatri in Italy, where he died in 1586.

The final member of the commission was Antonio Lilius, who represented his late brother's interest after presenting Aloysius's ideas in 1576—an event Gregory mentions in his 1582 bull by recalling that "a book was brought to us by our beloved son Antonio Lilio, doctor of arts and medicine, which his brother Aloysius had formerly written."

This "book," still in manuscript form, was easily the most important document in the entire reform process. Yet over the centuries it has disappeared without a trace. What survives is a short booklet issued by the commission, titled *Compendium novae rationis restituendi kalendarium*, "Compendium of the new rational for reforming the calendar." This is a synopsis of Lilius's plan sent out to various experts and important princes, monarchs, and prelates for comment.

The *Compendium* was also believed lost until the historian Gordon Moyer located not one but several copies in 1981—all printed in Rome in 1577. The booklet is a short quarto volume containing 24 pages, with a title page that prohibits the selling or reprinting of the book "under penalty of excommunication." All of the copies of the *Compendium* found by Moyer in archives in Florence, Siena, and Rome are attached to other short volumes critiquing Lilius's ideas— with some offering modified plans of their own.

The controversies that continued to swirl around talk of changing the calendar broke down along the familiar lines of science, theology, Church doctrine, and the practical impact of reform on the lives of people, governments, and the economy. By the 1570s, however, the emphasis was different, with the once potent theological concerns of God and time weighing in far less than debates about astronomic theory, Church cosmology, and how to mechanically come up with the best solution for fixing the calendar.

First on the list of contentious issues was the age-old conundrum: what is the true length of the year?

No one had yet come up with a method for determining the true year beyond a doubt—an issue still not entirely resolved today, given the variability of the earth's movements—even if the science of astronomy in the sixteenth century was slowly improving. Indeed, by the late 1570s it had become refined enough that Clavius and the commission could seriously consider whether the calendar should be changed to a system based on the actual motions of the earth (or the sun, if you were a follower of Ptolemy), instead of one that used a *mean value* of measurements. The latter was the method employed in both the Julian calendar, with its leap-year system, and by the Church's lunisolar calendar for determining Easter. Neither calendar had ever been linked to planetary theory; this had long appalled astronomers, who thought that the only way to create an error-free calendar was to drop the idea of a mean and to go on "real time," so to speak.

Clavius, for one, initially hoped to link up the reformed calendar as closely as possible to the true astronomic year. "I should think that in order to restore and keep account of astronomy it would be rather important to adopt the true motion," he wrote to a friend in Padua on 24 October 1580, "but these gentleman [of the commission] do not understand this for several reasons."

Lilius, however, argued in favor of a mean, insisting that astronomic theory remained too uncertain despite its advancements. He also believed that trying to devise a calendar based on planetary theory would be far too complicated for people who were not astronomers. What was needed, he said, was a mean calculated to be as close as possible to the true motions of the moon and the perceived motions of the sun.

Apparently the commission agreed, concluding that a calendar must be simple enough for all to understand and use, even if it is slightly off the true astronomic year—the challenge being to make the margin of error as small as possible. Even Clavius evidently came around and was persuaded to go with Lilius, since he later defended this choice after the reform was introduced.

That issue settled, the commission's next task was to decide which of the many measurements of the year they believed to be most reliable.

A half century earlier Copernicus had scratched his head and pondered the same question. He had decided that there were no good measurements for the tropical year, which seemed to him to speed up and slow down with no discernible pattern. This led him to rely on the more stable sidereal year in *De revolutionibus*. Calendar makers did not have this option, however, since their concern was with creating a "year" that matched the seasons, not the position of the earth in space—the two being slightly different, given that pesky phenomenon known as the precession of the equinoxes.

To understand this problem, and how it is possible to have two different kinds of years, first visualize the earth as a simple sphere or ball circling the sun. The sidereal year is the amount of time it takes for the earth to circle the sun relative to a fixed celestial object, such as a star; in other words, to reach the exact point in the orbit where it began.

SIDEREAL YEAR

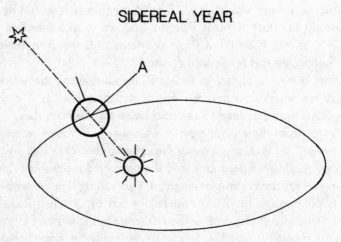

A = The starting and finishing point of the earth's orbit in space.

That's easy. Where it gets tough is when you realize that the earth not only spins around like a top—this is where we get our day and night—but also "tilts," its plane of rotation on its axis tilting relative to the plane of its orbit around the sun (the ecliptic).

To imagine this, think of the globe that sat in the front of your classroom in grammar school, with a line drawn around the fattest part: the equator. Without any tilt, the equator would always be the closest place on the earth to the sun, and we would have no seasons. But in fact the earth does tilt—so that in June the northern hemisphere is aligned with the sun on its ecliptical plane, when it is summer in

THE TROPICAL YEAR

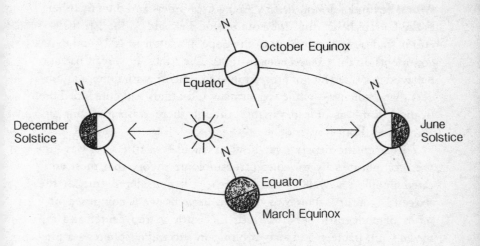

the north. Roughly six months later the earth tilts so that the southern hemisphere is aligned relative to the plane, making it summer in the south and winter in the north. In between the tilt brings the equator into perfect alignment with the ecliptic, marking the equinoxes that occur in March and September.

Hipparchus in Alexandria was one of the first astronomers to notice the difference between the two types of years when he took measurements of the year according to the equinoxes on his *skaphe*, from 141 to 127 B.C. He then must have compared this to the year as measured by the Egyptians, who for centuries had been measuring sidereal year rather than a tropical year. This is because they used as their time "ruler" the annual rise of the Dog Star, Sirius, catching it at the moment it could be seen crossing the peaked point of an obelisk.

Based on Hipparchus's observations, Claudius Ptolemy three centuries later proposed a simple formula for the precession, hypothesizing that the drift of the tropical year against the stars was fixed, and amounted to one degree per century.

By the time of the calendar commission this had been proven wrong beyond a doubt, first by Arab astronomers and then by others, as the Patriarch Ignatius, the commission's expert on the Islamic scientific tradition, pointed out to the pope in a letter in 1579 and in his comments on the *Compendium* in 1580. The Arabs, however, had also believed in a fixed rate of precession—coming up with different numbers than Ptolemy—while Copernicus and others had concluded the tropical year was indeed variable, though there was significant disagreement about how much.

This scientific debate over how to calculate a true year was further complicated by the ancient cosmologic theory that most educated people, as well as the Church, still considered true in the sixteenth century. This was that the heavens were composed of a series of concentric spheres, with the earth in the center and the moon, sun, planets, and stars orbiting in successive spheres—a precise and unchanging configuration that could not easily accommodate the possibility of a variable year, or of a starfield that seemed to be drifting slightly each year.

One explanation was that another, even higher sphere of stars might exist, or perhaps several more. This possibility created a great deal of muddle and confusion as traditional astronomers and ecclesiastics struggled mightily to make new and still sketchy data fit into their age-old conception of the universe.

The two astronomers on the calendar panel, Clavius and Danti, each had to convince himself that the year was in fact variable at a time when this was still controversial. For Danti the confirmation came when he took his measurement of the equinoxes in Florence in 1574 and 1575 and found that the length of the year differed from Ptolemy's measurements. The proof for Clavius came when he constructed a celestial globe for the Collegio Romano and calculated the rate of precession for the years between Copernicus's observations in 1525 and the year Clavius built his contraption in 1575. This was a change from his earlier blanket acceptance of all things Ptolemaic. Indeed, Clavius kept an open mind about the precession during the commission's debates, once referring the members to an unpublished essay by a certain Ricciardo Cervini, written in 1550, which argued that there was no precession at all, though Cervini failed to convince anyone.

Given the turmoil over the precession—and the larger controversy looming over Copernicus versus Ptolemy—Aloysius Lilius wisely ignored the entire issue in his solution. According to Clavius—our major source, along with the *Compendium,* for what Lilius was thinking, since Lilius's own manuscript is lost—the old physician opted simply to choose a value for the year based on what was then one of the more popular astronomical tables. These were the Alfonsine Tables, originally written in 1252 and updated over the years. They gave a mean tropical year of 365 days, 5 hours, 49 minutes, and 16 seconds. This was some 30 seconds slower than the true year, but still quite close. The mean value for the year used in the reform itself, which is our calendar year today, is slightly more accurate at 365 days, 5 hours, 49 minutes, and 12 seconds—a year that runs only 26 seconds slower than the true year.

This final mean for the Gregorian year allows us to summarize some key measurements, estimates, and guesses of the length of the tropical year taken over the centuries, most of which the commission had access to during the decade of their deliberation.

	Historical Lengths of the (Tropical) Year		
Year(s)	**Source**	**Measurement (days, hours, minutes, seconds)**	**Margin of Error from the Current Year**
Present	Atomic Clock	365d 5h 48m 46s★	None
141–127 B.C.	Hipparchus	365d 5h 55m	+6m 14s
45 B.C.	Julius Caesar	365d 6h	+11m 14s
A.D. 139	Ptolemy	365d 5h 55m 13s	+6m 27s

★This value is the standard length for our current era; determined in 1956, it is the mean year calculated for 1900.

Year(s)	Source	Measurement	Margin of Error
499	Aryabhata	365d 8h 36m 30s	+2h 47m 44s
882	al–Battani	365d 5h 48m 24s	−22s
c. 1100	Omar Khayyám	365d 5h 49m 12s	+26s
1252	Alfonsine Tables	365d 5h 49m 16s	+30s
c. 1440	Ulugh Beg	365d 5h 49m 15s	+29s
1543	Copernicus	365d 5h 49m 29s	+43s
1574–75	Ignazio Danti	365d 5h 48m	−46s
1582	Gregorian calendar	365d 5h 48m 20s	−26s

Once Lilius had decided on his mean year, he pondered the next crucial problem of reform: how to close the gap between Caesar's year and the "true" year. This meant comparing the Alfonsine year of 365 days, 5 hours, 49 minutes, and 16 seconds to the Julian year of 365 days, 6 hours. The Alfonsine runs short of the Julian by 10 minutes 44 seconds—equal to a day lost every 134 years.

Lilius seems to have tried different ideas to work this cumbersome measurement into a simple formula for dropping an appropriate number of leap days from the calendar. He rejected the long-standing proposal advocated by Bacon and others to drop a day roughly every 134 years. Instead Lilius took as his inspiration the simplicity of the Julian leap-year formula, with its easy-to-remember four-year rule, hoping to come up with a similarly convenient dictum to solve the Julian gap.

As the good doctor tinkered with various solutions shortly before his death, he discovered that the gap amounted to three days gained against the true year every 402 years (134 years × 3). This he rounded off to three days every 400 years, a more accessible number that became the basis for the leap-century rule—which drops three days from the calendar every four hundred years by canceling the leap year in three out of four century years. This formula, based on tables not entirely precise and a base number that is rounded off, ended up being remarkably accurate, running ahead of the seasons by only one day every 3,300 years.

Lilius also proposed two well-known options to recoup the days already lost due to the drift of the Julian calendar, which he thought should be cut by ten days to restore the equinox to the time of Nicaea. He suggested making up the days either by skipping 10 leap years over the course of 40 years, or—more radically—by removing ten days all at once.

The other big problem for Lilius and the calendar commission was repairing the Catholic lunar calendar used to determine Easter. Indeed, for the pope and other Christians the project of cinching up the solar calendar—and restoring the spring equinox to its proper place in the tropical year—was never an end in itself, but part of a religious fix required to restore the Feast of the Passion to its "proper" date.

Easter, of course, is supposed to fall on the first Sunday after the first full moon after the spring equinox—a seemingly straightforward formula, except for the ancient problem that the moon and sun do not match up in their respective years. To compensate for this Christian time reckoners had long used the 19-year Metonic cycle—which theoretically brought the sun and moon into sync because every 19 years of solar time equaled 235 lunar months.

Well, almost. In reality the moon's cycles run roughly an hour and a half behind the 19-year solar cycle, a mismatch that had been alarming computists and astronomers for some time.

Lilius calculated that the lunar-solar gap equals about 1 hour, 27.5 minutes, which meant that the moon was drifting against the Church's lunisolar calendar by a whole day every 312.7 years. By the 1570s this error had amounted to more than four complete days.

To halt this lunisolar mayhem, Lilius and the commission scrapped the old Metonic assumption that the phases of the moon, particularly the critical full moon, always matched up in the 19-year cycle with the solar year. Instead Lilius concentrated on trying to work out a new method for keeping the lunar calendar from sliding a day every 312.7 years.

Again, this was no easy task, given that 312.7 is hardly an easy number to divide into a Gregorian calendar of 365 days, 5 hours, 48 minutes, and 20 seconds. But once more Lilius came through, with a simple discovery that eight periods of 312.7 years equal almost 2,500 years—a number that can be divided almost perfectly into seven periods of 300 years plus one period of 400 years. This was Lilius's lunar solution: dropping one day from the lunar calendar every 300 years seven times, and then an additional eighth day after 400 years. For simplicity's sake Lilius and the commission again proposed making the corrections and dropping the days at the end of appropriate centuries.

Lilius's manuscript was initially received with some doubts and resistance, but soon it became the panel's lead proposal as Clavius and company studied it and sent it to various experts for comments. One so-called expert, Giovanni Carlo Ottavio Lauro, at one point seems to have tried to slow up the review process by taking Lilius's manuscript—and holding it for several months. Supposedly this was to make unspecified "corrections," though Lauro actually used the time to delay action so that he could finish his own proposal. His tactics so infuriated Lilius's supporters on the commission that they appealed directly to the pope, asking that the manuscript be returned—which it was—and the "chimeras" of Lauro be ignored.

Lilius's solution won out at last when the pope issued on January 5, 1578, the *Compendium* of the doctor's manuscript to universities, heads of state, and important prelates for their comments. The *Compendium* was sent rather than Lilius's much longer manuscript to save time now that calendar reform fever had struck Rome—or at least the small group of people who cared about such matters in the Eternal City. It also allowed the calendar committee to add its own remarks and amendments, which Clavius later says were minimal. The 20-page *Compendium* was written by the commission member from Spain, Pedro Chacón, presumably with input from Lilius's brother, Antonio.

After the publication, more comments poured into the commission. It received a vigorous response compared to past reform efforts, such as the one initiated earlier in the sixteenth century by Paul of Middelburg. This time the *Compendium* attracted dozens of letters, still preserved in the Vatican. Most simply gave their nod of approval; others contained comments, proposals, and counterproposals, some of them fascinating. The court mathematician for the duke of Savoy, Giovanni Battista Benedetti, made a number of suggestions in an April 1578 letter—including a calendar correction of 21 days, which would land the winter solstice on the first of January. Benedetti further proposed changing the length of the months to coincide with the presence of the sun in each of the 12 zodiac signs. Other commentators advocated various dates for the equinox and complained about using a mean for the length of the year. Some went to the trouble of publishing their alternative plans and circulating them, hoping to get a hearing with the commission and the pope.

Royalty also responded. For instance, King Philip II of Spain, in a short letter signed with a flamboyant *El Rey,* "The king," approved of the plan, but insisted that the equinox be kept on March 21—out of deference for Nicaea, but also for the practical reason that a great expense would be spared if the date did not have to be changed in mass books and breviaries.

The complaints of astronomers and other scientists would continue over the next several decades as the new calendar took hold. Most agreed with the technical side of the reform, including the Protestants Tycho Brahe and Johannes Kepler. Both found the reform scientifically sound and the best they had seen. Brahe from the beginning dated his letters using the new calendar, and Kepler in a posthumous article offered his arguments in the form of a dialogue between a Protestant chancellor, a Catholic preacher, and an expert mathematician. In the end he concluded that Easter, which was causing so much consternation among opponents and proponents of the calendar, "is a feast and not a planet." In 1613, Kepler argued in support of the reforms, but failed to persuade the Protestant sovereigns, a resistance that lasted until 1700. Even then Kepler's own Rudolphine Tables were substituted for the Gregorian values when determining Easter. In some years, this caused Germany to celebrate Easter on a different day than Catholics and other Protestants.

A great many astronomers found fault with the new calendar, including several mathematicians in Prague who refused to help the bishop there revise the calendar of feasts because they claimed to find the science unsound. Others disagreed, sometimes vehemently, for religious reasons. These included the Protestant astronomer Michael Maestlin (1550–1631), a professor at Tübingen in southern Germany and one of the teachers of Johannes Kepler. He insisted that the pope had no authority to institute such a reform, and also criticized Gregory for calling the new calendar "perpetual," because this denied the coming of the last Judgment. This argument was later refuted by another German defender of the calendar, who suggested that by Maestlin's reasoning people should also stop building houses.

Maestlin and others repeated criticisms that the reform should adhere more closely to the true movements of the sun (i.e., the earth) and moon. They complained about the methods used to determine Easter in the lunar reforms, worried over whether the equinox under the reform would always fall on March 21, and challenged the sources for the length of the year. Many astronomers and mathematicians—

including several assigned by monarchs and bishops to prepare the reforms for public dissemination—not only offered criticism but published their own solutions, sometimes side by side with the new calendar, to the confusion of anyone trying to understand the pope's reforms.

Other astronomers, led by Christopher Clavius, defended the new calendar. In 1595 he wrote a refutation of Maestlin, directed at the calendar's many critics, called *Novi calendarii Romani apologia, adversus Michaelem Maestlinum*—"Defense of the new Roman calendar, in reply to Michael Maestlin." He explained, among other things, why the commission adopted a system of mean rather than absolute motions.

Clavius also defended the use of a mean by pointing out that it was impossible for all Christians to celebrate Easter at exactly the same moment given the spread of Christians across several meridians. In 1606 Clavius answered his critics in the 800-page *Explicatio* (Explanation). In all, Clavius penned six treatises on the calendar, characteristically well-reasoned and scientifically sound documents that went a long way toward quieting the criticism and smoothing the way for reform in countries that initially hesitated to go along with the new calendar.

One of the most well known scholarly critics of the calendar was a bitter rival of Clavius, the French scholar and Calvinist Joseph Justus Scaliger (1540–1609). He found the reform littered with supposed errors and even stooped to name-calling, referring to Clavius as a "German fat-belly." But this did not keep Scaliger from later using the Gregorian system for his most famous project: creating a timeline of historical events according to the rules of astronomy. This was a monumental task, one that modernized the old medieval preoccupation with chronology and brought together all of the historical timelines and descriptions of events he could find. Indeed, he and Clavius were not so far apart in their respective tasks, the portly German setting out to align the calendar as closely as possible with the movements of the sun and moon, and Scaliger trying to get the past and future to correspond with a generally accepted standard. The year after the calendar reform Scaliger published *Opus de emendatione tempore* (1583), establishing chronology as a science.

Scaliger invented his own chronological calendar: the Julian day calendar, an ingeneous if complex system that does not use individual years at all, but a cycle of 7,980 astronomic years that counts a day at a time, with no fractional days, no mean year, and no leap years. He came up with his number by multiplying three chronologic cycles: an 18-year solar cycle, a 19-year lunar cycle, and the 15-year indiction period used by the Romans. All three cycles began together at the same moment at the start of his "Julian cycle," but would not converge again until the end. This was useful for anyone trying to create a uniform timeline, since the date from any one of the three base cycles could be translated into the two other cycles.

This may sound far too obtuse for the average person. However, Scaliger's calendar lives on today among astronomers, who do not need a calendar based on a mean of the tropical year but one that is astronomically exact. How else could one properly measure the time between, say, two appearances of the comet Hale-Bopp, or two pulses of a quasar? Scaliger began his Julian cycle at noon on January 1, 4713 B.C., which he based on calculations concerning Christ's birthdate.

The other great chronologist of the early modern era was Sir Isaac Newton (1642–1727), whose work in astronomy finally demolished what was left of the Ptolemaic school in planetary theory, and whose work on light, gravity, and mathematics launched modern physics. A man of many interests, Newton later in life became obsessed with properly dating the past. This included an elaborate attempt to correlate biblical events with those recorded in civilizations ranging from Assyria to Rome.

His astronomy and methods of dating long-ago events were brilliant, using recorded eclipses, the rate of drift in the precession of the equinoxes, and careful measurements of stars, equinoxes, comets, and novas. But his attempt to date myths and legends of dubious historic validity and his adamant piety about using the Bible to date events tainted his actual timeline. He insisted, for instance, that the world was created by God in 4004 B.C., as determined by Irish archbishop and student of the Scriptures James Ussher (1581–1656). He attempted to establish the entire timeline based on the voyage of Jason and the Argonauts in search of the Golden Fleece—an effort admirers called "masterly" and the work of "genius," but others dismissed as "no better than a sagacious Romance."

On September 14, 1580, the commission signed its official report to Gregory XIII, with Aloysius Lilius's solutions largely intact. They also added a clause to standardize New Year's Day on January 1, the date used by Julius Caesar.

Gregory enthusiastically approved the plan, which was set for implementation in October 1581—October being a month with few holy days. A final delay kept this from happening when the commission waited for a Flemish scholar named Adriaan van Zeelst to deliver promised improvements on Lilius's solution, though all he seems to have accomplished was to cause the postponement of the reform until 1582.

The bull itself was written in the fall of 1581, mostly by Pedro Chacón. On October 20, 1581, he sent a draft from Turino to Cardinal Sirleto in Rome. Chacón then died a few days later, leaving the final version of the bull to be written by commission member Vincenzo di Lauri. Sirleto also dispatched Antonio Lilius, Aloysius's brother, to work with the pope's aides on the final bull at Mondragone, Gregory's favorite villa outside of Rome.

On February 24, 1582, the 80-year-old Pope Gregory XIII sat down at a table that is still preserved at Mondragone and signed the bull that would make this the last year of Julius Caesar's calendar, at least for those staunchly Catholic countries still willing to accept a decree from the much-deflated authority of the Roman See.

On March 1 the text was posted at the doors of Saint Peter's, the chancellery of Rome, and other locations in the city. Printed together with the new perpetual calendar and the basics of the new system, copies were dispatched to every Catholic country through the papal nuncios as everything was prepared for a new calendric era, named for the pope who made the reform possible.

Gregory deserved this honor for the sheer bureaucratic feat of pushing through the reform when so many others had failed. Still, it seems unfair that the mysterious doctor who actually devised the reform didn't get some small measure of immortality for his troubles—perhaps a star named for him. Or, like Clavius, Copernicus, and Tycho Brahe, a crater on the moon.*

---

*Curiously, Clavius's crater is larger than those named after his more famous rivals. It also is the crater where the action took place in Stanley Kubrick's film *2001: A Space Odyssey*.

14

Ten Days Lost Forever

I grit my teeth, but my mind is always ten days ahead or ten days behind: it keeps muttering in my ears: "That adjustment concerns those not yet born."

—MONTAIGNE, 1588*

When bells chimed across Europe in the waning moments of October 4, 1582, the calendar did something it had not done since Julius Caesar's time: it jumped 10 days, at least in those countries that obeyed the pope's bull.

Anyone alive on what would have been October 5 instantly lost ten days of his or her life, according to Rome's new calendar. This genuinely upset people, who felt the days had somehow been stolen from them. In Frankfurt a mob rioted against the pope and mathematicians, who, they believed, had conspired to commit this theft. Others openly expressed their fear and unease at upsetting the saints they prayed to for everything from good crops to the afterlife in paradise. And everywhere people asked: What if the new days were wrong? What if the saints did not listen?

More mundane but practical were the sailors, muleteers, weavers, swordsmiths, and kings who worried about taxes not collected, wages not earned, and deadlines coming due 10 days early. Bankers scratched

---

*Montaigne actually wrote "eleven days," which was mistaken. His error suggests that even among intellectuals the reform was confusing.

their heads over how to calculate interest during a month only 21 days long, and local priests tried to explain to anxious parishioners that holy days were not the only dates bumped up; so were most other dates* from birthdays and wedding anniversaries to local fairs and civil ceremonies. Even the birthday of the pope had changed: from January 1, 1502 to January 11, 1502.

But the situation in October 1582 was far more confused than this. For only a scattering of countries actually enacted the hard-fought reform, with most people waking up on the morning after October 4 to no change at all: to the day that had always come next, October 5.

Had the Vatican issued its edict even a century earlier, it almost certainly would have been obeyed across a Europe then overwhelmingly Roman. But in 1582 the continent was a writhing, shifting patchwork of Protestants and Catholics; of kingdoms and dukedoms siding with Rome, or against it, or resting uneasily somewhere in the middle; of families and villages riven by loyalties to one faith or the other; of the Inquisition trying to root out not only Protestants but also Jews, Moslems, supposed witches, alleged heretics, and, in the case of Galileo, respected scholars who failed to kowtow to Rome's policies concerning the nature of the universe.

This was the era of the St. Bartholomew's Day Massacre in Paris; of the Catholic-backed attack of the Spanish Armada against Protestant England; of terror campaigns by Spanish troops against Dutch Protestants; of England's terrorism against Catholics in Ireland; of the Raid of Ruthven in Scotland, where Protestant nobles kidnapped King James VI and imprisoned him for 10 months; and of countless battles, sieges, and declarations of independence by Protestant cities and states in Germany and central Europe.

Set against this backdrop, Gregory's bull was a regrettably political document, a command from the pope as strident as anything produced by the pontific pen during these tumultuous days of the Counter-

---

*Some holidays did not shift ten days, but remained anchored in the new calendar on their original date, such as the Sabbath on Sunday and Christmas on December 25.

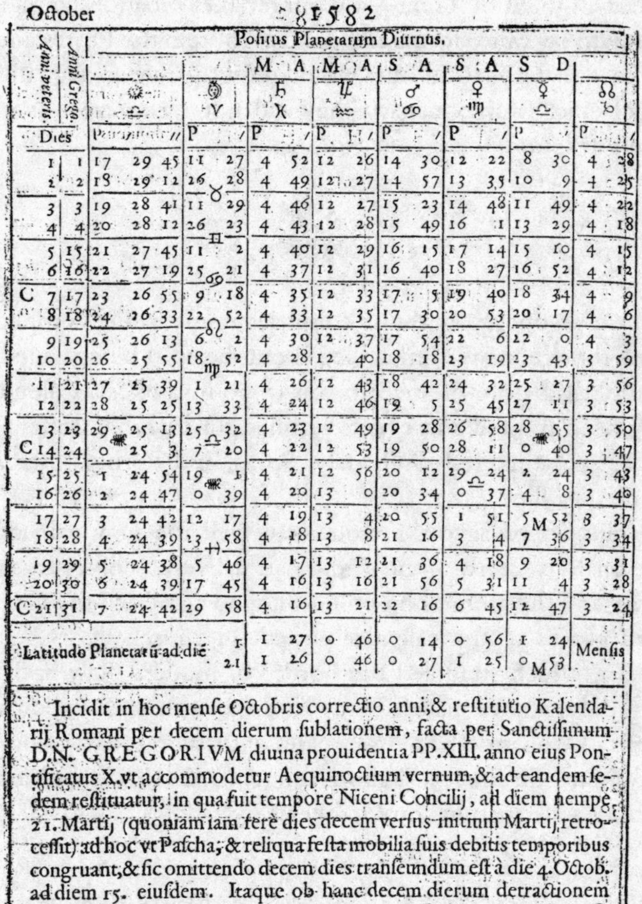

October 1582

Anni veteris	Anni Grego.	Positus Planetarum Diurnus																
		☉ ♎			☽ ♈		M A ♄ ♓		M A ♃ ♒		S A ♂ ♋		S A ♀ ♍		S D ☿ ♎		☊ ♑	
Dies		P	M	"	P	'	P	'	P	'	P	'	P	'	P	'	P	'
1	1	17	29	45	11	27	4	52	12	26	14	30	12	22	8	30	4	28
2	2	18	29	12	26 ♉	28	4	49	12	27	14	57	13	35	10	9	4	25
3	3	19	28	41	11	29	4	46	12	27	15	23	14	48	11	49	4	22
4	4	20	28	12	26 ♊	23	4	43	12	28	15	49	16	1	13	29	4	18
5	15	21	27	45	11	2	4	40	12	29	16	15	17	14	15	10	4	15
6	16	22	27	19	25 ♋	21	4	37	12	31	16	40	18	27	16	52	4	12
C 7	17	23	26	55	9	18	4	35	12	33	17	5	19	40	18	34	4	9
8	18	24	26	33	22 ♌	52	4	33	12	35	17	30	20	53	20	17	4	6
9	19	25	26	13	6	2	4	30	12	37	17	54	22	6	22	0	4	3
10	20	26	25	55	18 ♍	52	4	28	12	40	18	18	23	19	23	43	3	59
11	21	27	25	39	1	21	4	26	12	43	18	42	24	32	25	27	3	56
12	22	28	25	25	13	33	4	24	12	46	19	5	25	45	27	11	3	53
13	23	29 ♏	25	13	25 ♎	32	4	23	12	49	19	28	26	58	28 ♏	55	3	50
C 14	24	0	25	3	7	20	4	22	12	53	19	50	28	11	0	40	3	47
15	25	1	24	54	19 ♏	1	4	21	12	56	20	12	29 ♎	24	2	24	3	43
16	26	2	24	47	0	39	4	20	13	0	20	34	0	37	4	8	3	40
17	27	3	24	42	12	17	4	19	13	4	20	55	1	51	5 M	52	3	37
18	28	4	24	39	23 ♐	58	4	18	13	8	21	16	3	4	7	36	3	34
19	29	5	24	38	5	46	4	17	13	12	21	36	4	18	9	20	3	31
20	30	6	24	39	17	45	4	16	13	16	21	56	5	31	11	4	3	28
C 21	31	7	24	42	29	58	4	16	13	21	22	16	6	45	12	47	3	24
Latitudo Planetarũ ad diẽ						1	1	27	0	46	0	14	0	56	1	24	Mensis	
						21	1	26	0	46	0	27	1	25	0 M	53		

Incidit in hoc mense Octobris correctio anni, & restitutio Kalendarij Romani per decem dierum sublationem, facta per Sanctissimum D.N. GREGORIVM diuina prouidentia PP. XIII. anno eius Pontificatus X. vt accommodetur Aequinoctium vernum, & ad eandem sedem restituatur, in qua fuit tempore Niceni Concilij, ad diem nempe 21. Martij (quoniam iam ferè dies decem versus initium Martij retrocessit) ad hoc vt Pascha, & reliqua festa mobilia suis debitis temporibus congruant, & sic omittendo decem dies transeundum est à die 4. Octob. ad diem 15. eiusdem. Itaque ob hanc decem dierum detractionem litera Dominicalis mutatur hoc anno 1582. post diem 4. Octob. in C.

Page from an astronomical calendar for October 1582,
J. A. Magini, *Novae Ephemerides,* 1582

Reformation. Clavius and others insisted that the bull was never intended to be provocative against rival churches, either Protestant or Eastern Orthodox. But the mere fact that Gregory took his authority from the Council of Trent—a Counter-Reformation council called primarily to lay out reforms and policies to stem the Protestant tide— guaranteed that non-Catholics would resist the reform as an illegal and immoral edict from a papacy they did not recognize—even if the science *was* sound.

Of course, staunchly Catholic countries immediately complied with the bull, though many complained about the edict being issued a mere eight months before the reform was to go into effect. Calendars had already been printed and events planned for October 1582; all these now had to be altered. Still, Italy, Spain, and Portugal managed to make the deadline.

The nuncio of Savoy, for one, received the new calendar from Rome on May 28. By June 12 he had delivered copies to the duke of Savoy and the archbishop of Turin, who agreed to the change and ordered copies of the calendar posted on church doors across this dukedom, which straddled the border of modern Italy and France. One of these copies made its way to the British ambassador in Paris, Sir Henry Cobham (1538–1605). He sent it along to the English secretary of state, Sir Francis Walsingham (c. 1532–1590), with a dispatch on various state matters, on October 17, 1582:

> I send you . . . the Duke of Savoy's letters ratifying the Pope's new calendar, with the bull of the Pope's nuncio . . . The French king has likewise granted to this nuncio that the Pope's calendar shall be under his privilege printed and published.

Nations less secure in the Catholic fold, or in less of a hurry, did not immediately comply. France waited until December, when King Henry III ordered the change. Belgium and the Catholic states of the Netherlands also delayed until the end of 1582, with Flanders and parts of Belgium making the jump the day after December 21, which was followed by January 1. This meant skipping Christmas, which Thomas Stokes, an English merchant and spy living in Flanders, confirmed in a letter written on January 2, 1583 (Gregorian time) to Walsingham in London.

> Yesterday by proclamation from the Court, and proclaimed here in this town, "that yesterday" was appointed to be New Year's Day and to be the first of January; so they have lost Christmas Day here for this year—Bruges, the 23 December 1582, *stillo anglea*,★ and here they write the 2 January 1583.

Informed that some countries had not made the switch as scheduled, the pope on November 7 issued a reminder for noncomplying countries, ordering them to omit the 10 days between February 10 and 21, 1583. Gregory also chided these holdouts, ordering that "the method set out below shall be universally adopted, the whole unhindered by excuses or obstacles." By 1584 the remainder of Belgium had made the change. Hungary complied in 1587.

This covered most of Catholic Europe in the West, except for the Holy Roman Empire, which was itself a microcosm of greater Europe: a crazy quilt of rival kingdoms, duchies, fiefdoms, and city-states, some Catholic and some Protestant, and nominally lorded over by the Holy Roman Emperor. At the moment this was Rudolf II (1552–1612), king of Hungary and Bohemia. Rudolf is mentioned by name in the bull of 1582, with the pope making a personal appeal to him to carry out the reform. But he lacked the authority or the arms to impose much of anything beyond his own power base.

This left the individual German states largely on their own. In October 1583 Bavaria and Austria converted. So did Wurzburg, Münster, and Mainz in November of that year, though each dropped a different set of 10 days. The Catholic cantons of Switzerland changed on January 12–22, 1584; most other German Catholic states, along with Bohemia and Moravia, became Gregorian by the end of 1584.

★"English style."

Protestants in Germany and elsewhere rejected the reform, often with great bitterness and passion. James Heerbrand, a professor of theology in the German city of Tübingen, accused Gregory—whom he called *Gregorius calendarifex,* "Gregory the calendar maker"—of being the "Roman Antichrist" and his calendar a Trojan horse designed to trick real Christians into worshiping on the incorrect holy days.

> We do not recognise this Lycurgus (or rather Draco, whose laws were said to be written in blood), this calendar-maker, just as we do not hear the shepherd of the flock of the Lord, but a howling wolf. . . . All his loathe-some and abominable errors, his sacriligious and idol-worshipping practices, his vicious, perverse and impious dogmas that are condemned by the word of God . . . these little by little he will once more insert into our churches.

Heerbrand vilified the new calendar as an extension of the Council of Trent and accused the pope of promulgating a religious change rather than a civil one. His advice: act as shepherds against the "slob-bering wolf that threatens your flock," and "stand firm in that liberty of yours, and fight for it as befits strong athletes and soldiers of Christ."

Other Protestant intellectuals argued that the pope's calendar was against nature, with one tract insisting that farmers no longer knew when to till their fields and that birds were confused about when to sing and when they should fly away. Another pamphlet coauthored by the anti-Gregorian astronomer Michael Maestlin frightened farmers in Bohemia and elsewhere by proclaiming that the pope really was stealing 10 days of everyone's life. Catholics countered with absurdities of their own, insisting that in Gorizia, Italy, a nut tree had responded to the papal reform by blossoming 10 days early. Other Protestants agreed with Martin Luther's reaction when he heard about Catholic reforms: how earlier in the sixteenth century that civil authorities, not popes, ought to be in charge of how society measured time. Still others insisted that the Julian calendar had been chosen by God and should not be altered by popes or kings—a position that the Catholic Church itself had of course taken for centuries, using the same argument to block reform of the calendar.

For the people of Germany and elsewhere this jumble meant that people now had to cope with *two* calendars: the Julian in the Protestant countries and the Gregorian in the Catholic ones—soon to be known as the "old style" and the "new style." or *O.S.* and *N.S.* for short. It also meant that someone leaving, say, Catholic Regensburg in Bavaria on January 1 would arrive in Lutheran Nuremberg, some fifty miles away, on December 21 *the previous year.** Worse, Christian holidays including Easter now fell on different days in a reprise of what Venerable Bede had complained about in distant Northumbria during the Dark Ages. "It is said that the confusion in those days was such that Easter was sometimes kept twice in one year," wrote Bede in 732, a sentiment now reemerging eight hundred years later, but this time across the length of Latin Europe.

Later, in 1700, Protestants in Germany and Denmark adopted most of the Gregorian reforms, including the removal of 10 days and the century leap-year rule. But they deviated on how they calculated Easter, which ended up producing an Easter date identical to the Catholics' except in certain years—such as 1724 and 1744, when Catholics and Protestants celebrated the Paschal feast on different days. In 1775 Frederick the Great finally suppressed the Protestant Easter calendar, after which the complete Gregorian calendar ruled in Germany.

Most confusing of all was Sweden, which adopted the German Protestants' new Easter calculation but did not remove 10 days from

---

*Regensburg accepted the Gregorian calendar in 1583, Nuremberg in 1699.

their solar calendar. Instead they dropped a single day in 1700, conforming to the Gregorian century leap-year formula that was being followed by all reformed countries that year. This left the Swedes with a different calendar than anyone else: 10 days out of sync with the Gregorian, but also 1 day off from the Julian. In 1712 they reverted back to the Julian calendar by adding an extra leap day, February 30. Only in 1753 did the Swedes at last adopt the Gregorian year.

The Eastern Orthodox Church also rejected the reform, a last-minute effort by Rome to include them having proven unable to undo centuries of enmity. If anything, the old hostility had grown worse since the fall of Byzantine to the Turks more than a century earlier—a defeat made more bitter to many Eastern Christians because they believed that the West had stood by and did nothing to help.

Since the fall of Constantinople the churches of the East had been thrust into a minority position within a powerful Moslem empire, though they continued to operate in their chief cities. But the central authority linked to the old Greek empire was gone, leaving local churches in Constantinople, Alexandria, Antioch, and elsewhere to fend for themselves in the sometimes hostile environment of Ottoman rule.

Even to send an official delegation from Rome to Orthodox leaders was a risky proposition in the 1570s and 1580s, given the Turks' sensitivity to anything that would encourage an alliance between Christians in the East and West. They were particularly edgy in the wake of military setbacks on their European frontier with the West in the Balkans, and after the 1571 Battle of Lepanto in the Gulf of Patras. There a combined Spanish and Italian fleet had decisively defeated the Turkish navy and ended the Ottoman domination of the Eastern Mediterranean sea lanes.

So the pope sent his calendar emissary east under cover, dispatching in May of 1582 a certain Livio Cellini in the guise of a trader traveling with a state delegation to Constantinople from Venice, which had a trade treaty with the Turks. Arriving on May 27, Cellini went the next day to visit Jeremiah II Tranos, the patriarch of Constantinople.

This was not the first contact with representatives of the Greek Church regarding the calendar. Gregory's commission had earlier sought input from the Orthodox bishop in Venice and had worked closely with the Syrian member of their panel, Patriarch Ignatius, hoping to assuage the Greeks and to bring them along. The commission seriously talked about inviting representatives from the East to attend reform discussions in Rome. But this was rejected in 1581 by Clavius and the others. They feared it would delay the reform and perhaps kill its chances, in part because so much depended on Gregory himself, who at age 80 could not be expected to live forever.

In Constantinople, Jeremiah was sympathetic to the reform, though he explained to Cellini that many of the other Eastern churches would be openly hostile to anything that came from Rome. Still, the patriarch made an effort to persuade the others. This was wrecked, however, when the news arrived that Gregory had unilaterally issued his bull the previous February. A synod held in Constantinople in November 1582 harshly condemned the reform as being against tradition, the Scriptures, the councils, and the wishes of the founders of the Church. They also chided the entire process of reform by decree from Rome as a vanity of the pope.

The Eastern churches remained entirely opposed to the Gregorian calendar until just after the First World War, when a congress of Orthodox churches met in 1923 in Constantinople. One of the items on the agenda was the "new" calendar, which was not formally adopted by the congress.* Since 1923, several individual churches in the East have adopted portions of the new calendar, including the switch to the Gregorian solar year. They retained the old system for calculating Easter, however, and to this day celebrate Christ's resurrection on a different day than Christians in the West.

---

*The conferees in this tumultuous gathering agreed on nothing else either.

These partially reformed churches include those of Constantinople, Alexandria, Antioch, Greece, Cyprus, Romania, Poland, and most recently Bulgaria, which made the conversion in 1968. The churches in Jerusalem, Russia, and Serbia and the monasteries on Mt. Athos in Greece continue to adhere entirely to Caesar's calendar, which now runs 13 days behind the Gregorian calendar. Small bands of "old calendarists"—called *Palaiomerologitai*—continue to hold out in Greece, following the Julian calendar despite being excommunicated by their church for failing to abide by the reforms.★ Only the Orthodox Church of Finland, with about 60,000 members in this overwhelmingly Lutheran nation, has switched entirely to the Gregorian calendar, including Easter.

How the average person reacted to the new calendar in the 1580s can be glimpsed only in bits and pieces, since Europe as yet had no Rome daily newspaper, no *Paris Match,* and no *London Times*. And few people kept diaries and journals—a practice that would have to wait for the newly literate upper middle class that began to appear late in the next century, and the new consciousness of time and individual worth during the Enlightenment that made people believe that what happened to them was worth writing down.

For people living in regions that adopted the new one, the change probably made little difference anyway from a practical standpoint, once a villager in Tuscany or the Loire Valley got over the shock of holy days changing and dealt with any lingering fears of 10 days lost. In 1582 most people still led very insular lives compared to today, seldom straying from their villages and fields. A few more were educated than in the time of Bacon and certainly more than in the era of Charlemagne,

---

★The monasteries on Mt. Athos are allowed to retain the Julian calendar because they are part of the Church of Constantinople, which has tolerated their position.

and most had enough food. Yet daily life in 1582 remained much as it had for centuries: filled with hard labor during planting and harvest, but with comparatively little to do the rest of the year; with moments of pleasure interspersed with the age-old perils of disease, war, famine, and—for some—religious persecution.

Time continued to intrude ever more urgently into the ancient cycle of life and death, with the continued spread of clocks and bells and a growing time consciousness of labor, trade, taxes, contracts, and so forth, which few people could avoid by 1582. This meant that most Europeans living in Gregorian countries would have heard about the change sooner or later, if only because they now prayed to saints on different days. Yet a certain timelessness would persist for some until well into the twentieth century, and remains even today in scattered places.

For those people who lived in a village that went Gregorian when the next village over stayed Julian, the calendar change would have been more obvious. For instance, how would the 10-day gap affect our person trekking over the mountains from noncomplying Nuremberg to the newly Gregorian Regensburg? If he was a muleteer driving a caravan loaded with charcoal, was he considered 10 days late? And would a woman married on June 10 in Regensburg be unmarried in Nuremberg the same day, which was their June 1?

Most people undoubtedly reacted to such oddities and inconveniences with a grumble and a shrug. Dates and systems of dating had been scrambled for so long, with competing saint's days, different New Years, and names for days that people were probably used to having to think simultaneously in more than one system. This is probably what our muleteer would have done. Anyway, he would not have fretted as much as we would today over such discrepancies, for the simple reason that few people in the 1580s cared about following the exact time. Most clocks still kept time only to the quarter hour. And no one had a train to catch at exactly 5:02 P.M., or a favorite television show they did not want to miss.

In Moravia a local saga about the change suggests the sort of thinking and talking that was going on by regular people concerning the calendar. In this tale a simple innkeeper named Bartholomaeus tries to understand the move from old to new. Presented as a tale of good versus evil, Bartholomaeus is advised by a priest and a devil. Moravia being a Catholic country, one can guess which advisor was which—and the outcome.

Throughout the great Gregorian time switch, few people probably focused on the role of science, not realizing that this shift was one of the first instances in the early modern age where a change affecting almost everyone was compelled less by religion than by a new respect for scientific accuracy—in this case, for getting the time right.

Nowhere was the turmoil over the calendar more evident than in England, which in the early 1580s was a country of three or four million people just beginning its rapid rise to the status of a world economic and military power. For now, however, this small island kingdom was weak and isolated, ruled by a Protestant queen who had spent her entire reign trying to protect herself and her realm against the great Catholic powers of the day, particularly Spain.

Imprisoned in 1554 by her Catholic sister, Queen Mary, who suspected her involvement in a Protestant plot, and having survived several Catholic conspiracies including an attempted assassination, the wily Elizabeth I (1533–1603) in 1582 was as embroiled as ever in trying to fend off her enemies. This makes it all the more surprising that when she heard about the pope's bull she did not reject it outright. Instead she asked her friend and advisor John Dee (1527–1608) to study and comment on the reforms.

A scientist, astrologer, and longtime confidant of Elizabeth, Dee was a fascinating character, a man who in many ways epitomized the Elizabethan era of Francis Bacon, William Shakespeare, Sir Francis

Drake, and Sir Walter Raleigh—a period of unusual impetuosity, wit, exploration, entrepreneurship, conquest, and an anything-goes mentality. Dee himself was a graduate of Cambridge, an editor of Euclid, an expert on navigational instruments, and an astrologer and conjurer who discoursed on everything from the nature of angels to Copernican theory. The son of Henry VIII's chief carver and manager of the royal kitchen, Dee also had traveled widely as a young man, furthering his studies of astronomy and cosmology in Belgium and lecturing to large crowds at the university at Reims. A minor sensation on the continent, Dee was offered positions at the courts of the French king and of Ivan the Terrible in Russia.

Instead he returned to England in 1551 to became an intellectual at the court of Queen Mary, soon switching his allegiance to the queen's half sister, Elizabeth. At one point Dee faced a charge of treason for supporting Elizabeth but was acquitted. This earned him the devotion of Elizabeth, who, when she became queen after Mary's death in 1558, asked Dee for astrological advice on the best date for her coronation. Later she called him simply "hyr Philosopher."

Dee took his work on the calendar very seriously. In 1582 he penned a long, passionate treatise in support of the reform, titled:

> A playne discourse and humble advise for our gratious Queene Elizabeth, her most Excellent Majestie to peruse and consider, as concerning the needful reformation of the vulgar Kalendar for the civile yeres and daies accompting or verifying, according to the tyme truely spent.

Dee also included on the flyleaves a little ditty intended to flatter Elizabeth and to not-so-modestly point up his own effort with Elizabeth as on a par with those of Sosigenes and Caesar:

> *As Caesar and Sosigenes*
> *The Vulgar Kalendar did make,*
> *So Caesar's pere, our true Empress.*
> *To Dee this work she did betake.*

This work started with a simple introduction to the problem, and then includes a circular timeline, or dial, around which Dee wrote the great names in the history of the calendar: Caesar, Hipparchus, Ptolemy, Bacon, and others. He then plunged into an analysis of the science behind Aloysius Lilius's reforms, particularly the length of the year. Consulting Copernicus's *De revolutionibus,* Ptolemy's *Almagest,* and Erasmus Reinhold's Prutenic Tables, he satisfied himself that the work done by Lilius and the calendar commission in Rome was sound and that the reforms were a sensible solution—with one exception.

Not being a Catholic, Dee had trouble with the dating of the calendar correction back to the Council of Nicaea. He vigorously argued for a restoration back to the time of Christ, which meant dropping 11 days, not 10. Dee later relented with great regret, supporting a 10-day drop to conform with the rest of Europe. He also wrote out a proposed calendar for 1583 with the 10 days deleted, advocating a less traumatic plan than the pope's elimination of the days all at once. Under this calendar England would have dropped three days in May, one in June, and three each in July and August, at times that avoided important days and holidays.

Once finished, Dee sent his treatise and sample calendar to the man who apparently headed up the queen's official commission looking into the matter, Lord Burghley, the lord treasurer of England. Dee began the report with another poem, emphasizing in exceptionally bad verse that the point of this reform was to be true to science:

> *At large, in brief, in midell wise,*
> *I humbly give the playne Advise*
> *For word of tyme, the Tyme Untrew*
> *If I have myst, Command anew*
> *Your Honor may: So shall you see,*
> *That Love of Truth, doth govern me.*

Burghley read the *Discourse* and then consulted with three other intellectual advisors to the queen: the mathematician Thomas Digges

(d. 1595), Sir Henry Savile (1549–1622), and a Mr. Chambers. These experts added their approvals and referred the matter to the Queen's chief councilors. They too approved the plan, as did the Queen, who set a date for implementation in May 1583.

Before they could move, however, one hurdle remained: the approval of the archbishop of Canterbury, Edmund Grindal (c. 1519–1583), and key bishops in the Church of England. To secure this, Walsingham, the secretary of state, dispatched a letter on March 18, 1583,★ asking the archbishop to confer with his bishops and return his response "with all convenient speed, for that it is meant the said callendar whall be published by proclamation before the first of May next." Walsingham followed up on this just 11 days later, on March 29, with another note urging Grindal to respond quickly. He suggested that the queen herself was anxious to receive his official nod. "Her majesty doth now find some fault that [she] doth yet hear nothing of the reports thereof that she looked to have received your Grace," wrote Walsingham.

Nothing could have been plainer, except for one problem—Archbishop Grindal said no.

Part of his obstinacy was a long-standing feud between him and the queen that undoubtedly would have led to his forced resignation had he not died that very year. But more than this was the aged archbishop's deep distrust of Rome, a stance that represented a strong current in the Anglican Church and in an English society in the 1580s that was proud to the point of xenophobia about their new religion, their hatred of Spain and the Catholics, and their love for their queen.

The savvy Elizabeth understood this—which makes her support of the measure all the more perplexing. Possibly she was simply succumbing to the eagerness of the intellectual circle at her court, the poets, scientists, adventurers, and philosophers who spent their time delighting one another—and Elizabeth—with their wit, wisdom, and

---

★That is, on March 28, 1583, according to the new Gregorian calendar. Because England not only was on the Julian calendar but started their year on March 25.

earthy good sense, when they weren't intriguing against the queen's enemies at home and abroad. But Elizabeth was also a pragmatist, a consummate political tightrope walker with an uncanny ability to fend off enemies and impassion loyalists.

Apparently she agreed with "hyr Philosopher" that the reform was good science. She may also have been convinced by Dee's assessment that the reform had a British connection through Roger Bacon. Undoubtedly she had a political motive, though what it was is unclear. It may have been part of her delicate game of tacking toward and away from Spain in these years leading up to the attempted invasion by the armada. Or possibly it was an attempt to enforce her will on the archbishop in their long-standing tug-of-war.

Whatever it was, Grindal dispatched his reply on April 4, including comments from key bishops and a "godly learned in the mathematicalls." The gist of the letter to Walsingham was a masterful strategy that avoided saying no outright. Instead, Grindal asked for a delay by insisting that a change this sweeping should be discussed in a general council of all Christians, such as the one convened in Nicaea by Constantine.

> After our hearty commendations unto your honour, may it please you to understand, that upon receipt of your letters in Her Majesty's name, and the view of Mr. Dee's resolutions . . . we have upon good conference and deliberation . . . that we love not to deale with or in anye wise to admit it, before mature and deliberate consultation had, nott only with our principall assemblie of the clergie and convocation of this realme, but also with other reformed Churches which profess the same religion as we doe, without whose consent if we should herein proceed we should offer juste occasion of schisme, and so by allowinge, though not openly yet indirectly, the Pope's dewyse and the [Trent] counsayle, [cause] some to swerve from all other Churches of our profession.

Grindal thus deflected the pressure exerted on him personally by insisting on a meeting that would never happen, even among the fractious Protestants. Grindal also argued that the Church of England could not under the rule of Scripture or God endorse an edict from a papacy that "all the reformed Churches in Europe for the most part doe hold and affirme . . . is Antichrist." In a long list of reasons why the calendar should not be reformed, Grindal and his bishops also reminded Walsingham that it would be particularly loathsome to accept an edict issued as a bull, since it was this same instrument of the pope's authority that had excommunicated Elizabeth in 1570.

Dee countered by saying that the new calendar had nothing to do with the pope, that it was astronomy that dictated the change. He pointed out the need of a rising maritime power to conform with its trading partners on the continent in something so basic as dates. But the matter was dropped after an abortive attempt to pass it in Parliament in 1584 (Old Style)—titled "An Act giving Her Majesty authority to alter and make new a Calendar according to the Calendar used in other countries." This bill was introduced on March 16 and possibly reread on March 18. It then disappears along with all efforts to change the calendar, for reasons that are not recorded. Possibly the queen and her advisors simply dropped the matter so as not to push the issue of the state versus the church as the possibility of war with Spain increased.

Shortly after the calendar debate ended, Dee left the English court for eastern Europe, traveling with his family and a "spirit medium" named Edward Kelley. In Bohemia he continued his intellectual pursuits and got involved with several dubious affairs involving astrology and angel readings with Kelley at the court in Prague. For the rest of his life Dee argued for the adoption of the new calendar in England, though after the attempted invasion by Spain in 1588—launched with the support of the pope—the revulsion for all things Roman made any reform impossible.

It would be another 170 years before Britain finally adopted the Gregorian calendar; it was one of the last major European countries to do so. This was despite serious reform attempts in 1645 and 1699, both blocked by a still strident Church of England and by Puritans taking the line that the "old stile" calendar was the true style of God.

But as Britain became a major international military and economic power, the inconvenience of the "old stile" and "new stile" became increasingly a nuisance for businessmen and an embarrassment for anyone with connections on the continent. "The English mob preferred their calendar to disagree with the Sun than to agree with the Pope," chided Voltaire. And in Latin someone wrote a ditty reprinted in a pro-reform tract in 1656:

> Cur Anni errorem non corrigit Anglia notum,
> Cum faciant alii; cerncre nemo potest.
>
> *Why England doth not th'years known error mend,*
> *When all else do; no Man can comprehend.*

Still, over the years most people in Britain and, as time went on, in its colonies seemed to take the inconveniences in stride, with overseas letters dated with two dates—O.S. and N.S. Over the years the English even seem to have developed a certain amount of pride (or arrogance) in being different—something akin to Americans' turning up their noses at the metric system today.

And here the matter stood until one spring day in 1750, on the tenth of May, when a stodgy earl named George Parker (1697–1764) stood up to deliver to the Royal Society an address with a seemingly deadly dull title: "Remarks upon the Solar and the Lunar Years, the Cycle of 19 years, commonly called The Golden Number, the Epact, And a Method of Finding the Time of Easter, as it is now observed in most Parts of Europe." Parker, an amateur astronomer well connected with the Newtonian circle in Greenwich and London, started his talk by updating just how far off the Julian year had drifted against the true year since Caesar's time—and since the Gregorian reform. As a point of reference, he used what was then perhaps the most accurate measurement of the year ever—365 days, 5 hours, 48 minutes, and 55 seconds—calculated by the late royal astronomer Edmund Halley (1656–1742), the man who gave his name to Halley's Comet.

"We do as yet in England follow the Julian Account or the Old Style in the Civil Year," Parker noted toward the end of his mostly technical talk, "as also the Old Method of finding those Moons upon which Easter depends: Both of which have been shewn to be very erroneous."

Most likely, the earl's speech would have gone unnoticed except for one member of the audience: the recently retired secretary of state Philip Dormer Stanhope (1694–1773), the earl of Chesterfield. Famous for his wit and sophistication, and for his sagacious letters to his son and godson, the 56-year-old Stanhope was for some reason fired up by the old earl's speech and launched an effort to push for reform at last in Britain.

Still an important member of the Whig Party and a prominent intellectual during this golden age of the drawing room, Stanhope first consulted with mathematicians and astronomers. He then took his cause to the leaders of his party; starting with his longtime political colleague Thomas Pelham (1693–1768), the secretary of state and future prime minister.

Pelham initially gave the idea a cool reception, as Stanhope later recounted. "He was allarmed at so bold an undertaking," Stanhope wrote, "and conjured me, not to stir matters that had been long quiet, adding that he did not love new fangled things." In another account of this meeting, the editor of Pelham's memoirs, William Coxe, agrees that the future prime minister was none too thrilled. "The noble secretary was too deeply impressed with the favorite maxim of Sir Robert Walpole," wrote Coxe, "*tranquilla non movere* [do not disturb things at rest], to relish a proposal, which was likely to shock the civil and religious prejudices of the people."

To overcome this inertia Stanhope set out to embarrass his country-men into change, pointing out to everyone who would listen what he later wrote in a letter to his son: that other than England, Russia and Sweden remained unreformed. "It was not, in my opinion very hon-orable for England to remain in a gross and avowed error, especially in such company, the inconveniency of it was likewise felt by all those who had foreign correspondences, whether political or mercantile." Stanhope also took his proposal to a medium that was unavailable to Christopher Clavius or to John Dee in the 1580s: the popular press. He penned a number of amusing and informative articles under a pseudonym in an eighteenth-century London periodical, *The World*. The affable earl also talked up the change in fashionable London town-houses, parliamentary antechambers, smoking rooms, and estates.

Eventually winning Pelham's approval and that of other senior gov-ernment ministers, Stanhope in 1751 introduced a bill for reforming the calendar in Parliament: "An Act for Regulating the Commence-ment of the Year, and for Correcting the Calendar now in Use." In a letter to his son, he writes: "I had brought a bill into the House of Lords for correcting and reforming our present calendar. . . . It was notorious, that the Julian calendar was erroneous, and had overcharged the solar year with eleven days." He then described his preparations for the bill and his presentation, in part as a lesson for his son on how to comport oneself in presenting a complicated matter in public.

I determined, therefore, to attempt the reformation; I consulted the best lawyers and the most skillful astronomers, and we cooked up a bill for that purpose. But then my difficulty began: I was to bring in this bill, which was necessarily composed of law jargon and astronomical calculations, to both of which I am an utter stranger. However, it was absolutely necessary to make the House of Lords think that I knew something of the matter; and also to make them believe that they knew something of it themselves, which they do not. For my own part, I could just as soon have talked Celtic or Sclavonian to them, as astronomy, and could have understood me full as well: so I resolved . . . to please instead of informing them. I gave them, therefore, only an historical account of calendars, from the Egyptian down to the Gregorian, amusing them now and then with little episodes. . . . They thought I was informed, because I pleased them; and many of them said, that I had made the whole very clear to them; when, God knows, I had not even attempted it.

Stanhope had laid his groundwork well. The bill sailed through the usual three readings and was passed on May 17 with a unanimous vote and approved by King George II on the 22nd, after which Stanhope quipped that it was his "style that carried the House through this difficult subject" and not the content of what he said concerning the mathematics and science, which "he himself could not understand."

The act itself ordered 11 days expunged from the calendar in Great Britain and in its colonies, with Wednesday, September 2, 1752, followed by Thursday, September 14. The 11th day was added because in 1700 the Gregorians, according to Lilius's century leap-year rule, had not observed a leap year and did not add a day. This meant that the Julian calendar, which *had* added a day, was 24 more hours out of step. The act also mandated that in the future the calendar year and Easter be observed according to the Gregorian system, and that the year would begin on January 1 in England instead of March 25.

Stanhope and Parliament took pains to legislate details of the changeover to minimize problems with banking, contracts, holidays, and matters public and private. For instance, the act explains that all court dates, holidays, "Meetings and Assemblies of any Bodies Politick or Corporate," elections, and all official obligations according to "Law, Statute, Charter, Custom or Usage" shall be "computed according to the said new method of numbering and reckoning the Days of the Calendar as aforesaid, that is to say, 11 Days sooner than the respective Days whereon the same are now holden and kept."

Similar provisions applied to markets, fairs, and marts, "whether for the sale of Goods or Cattle, or for the hiring of Servants, or for any other Purpose," and to rents, usages of property, contracts, "the Delivery of such Goods and Chattles, Wares, and Merchandize," with the Act ordering that no one was to pay wages or count or pay interest for the 11 lost days. Even those who happened to be turning 21 years of age between September 3 and 13, 1752, O.S.—this was the legal age of majority in Britain—did not get a break. Nor did soldiers about to be discharged from the army, indentured servants at the end of their contracts, or criminals about to be released from jail. They all had to wait the proper number of "natural days" that would have occurred under the old calendar.

During the months between the vote and enactment the government acquired an unlikely ally in the Church of England, which had finally lined up in favor of reform and adopted a slogan: "The New Style the True Style." This became the motto of preachers across Britain, who added an appeal to patriotism by repeating John Dee's assertion that Roger Bacon, an Englishman, was among the first to call for reform some five hundred years earlier.

The act was also disseminated in the *London Gazette* and in other newspapers and almanacs. For instance, *The Ladies Diary, or Woman's Almanack*, published in London, offered a detailed explanation of the change on their cover and in their calendar for the month of September (see pages 283–284).

Still, many people in Britain reacted with dismay when September actually rolled around—and, in some cases, with anger at the confusion over 11 days lost. William Coxe, editor of Pelham's memoirs, summarized the response:

> In practice . . . this innovation was strongly opposed, even among the higher classes of society. Many landholders, tenants, and merchants, were apprehensive of difficulties, in regard to rents, leases, bills of exchange, and debts, dependent on periods fixed by the Old Style. . . . Greater difficulty was, however, found in appeasing the clamour of the people against the supposed profaneness, of changing the saints' days in the Calendar, and altering the time of all the immoveable feasts.

In London and elsewhere mobs collected in the streets and shouted, "Give us back our 11 days." This became a campaign slogan in 1754

The Ladies Diary, Cover and (on the next page) Month of September, London, 1752.

1752 September hath only XIX Days in this Year.

First Quarter, 15th, 2 After.	*Sun* enters ♎ 22d. 5h. 1m. 11s.
Full Moon, 23d, at Noon.	E. T. Equat. + 7ʹ. 41ʺth. 22d
Laſt Quarter, 30th, 1 After.	5h. 8m. 52ſ. Ap. T. *Sun*'s mean
	An. 2ſ. 23°. 23ʹ. 44ʺ. 59ʹʹʹ.

1	T	*Giles, Abbat & Conf.　Sun faſter than Year* 3ʹ 55ʺ	8 A 12
2	W	*London* Burnt, 1666.　*Sun* riſes 5. 37, ſets 6, 22.	8　49

BY 365 *Days*, 6 *Hours, the mean* Julian *Year, being long reckon'd for* 365 d. 5h. 48m. 54ſ. 41th. 27 fourths, *the Year by the Sun, according to Dr. Halley,* (See Palladium 1750, p. 53.) *The Account of Time has each Year run a-head of Time by the Sun* 11m. 5ſ. 18th. 33 fourths, *or* 44m. 21ſ. 14th. 12 fourths, *every* 4 *Years, and conſequently* 3d. 1h. 55m. 23ſ. 40 thirds *in* 400 *Years : And ſo from the Council of Nice, when the Kalendar was ſettled, in the Year* 325, *to this preſent Year* 1752, *being* 1427 *Years, the Time by Account is forward of that by the Sun* 10 d. 23 h. 43 m. *and therefore* 11 *Days is left out of Account, in this Month, as the moſt convenient, for reducing the Kalendar or Year to its firſt eſtabliſh'd Order. And for keeping the ſhorteſt and longeſt Days (or the Solſtices) and alſo the Days of* 12 h. *long (or the Equinoxes) on the ſame nominal Days of the Month for the future, it is ordain'd by Act of Parliament, that every fourth hundred Year is to conſiſt of* 366 *Days as uſual, but all other whole hundred Years of* 365 *Days only : The Years between: which whole hundreds to be common and* Biſſextile *as formerly, and the Date of the Year henceforward to begin on the firſt of* January.

14	T	*Holy* Croſs *Day, Holy-Rood, or Exalt. of the Croſs*	9	33
15	F	Day 12 hours 20 minutes long	10	24
16	S	Day decreaſed 4 hours 14 min.	11	18
17	A	15 Sund. after Trinity.　*Lambert,* Biſh. and Mar.	Morn.	
18	M	*Planetary Hour by Day* 62 *minutes*	0	18
19	T	*Birth of the Virg.* Mary. *Dunſtan.　Sun* 6ʹ 22ʺ *faſt*	1	22
20	W	*Faſt.*	2	27
21	T	St. MATTHEW, Apoſtle, Evangeliſt, and Martyr	3	35
22	F	*Equal Day and Night in all the Habit. World.　Sun*	4	45
23	S	*Sun* 7ʹ 44ʺ *too faſt　　[truly riſes and ſets at 6*	☽ riſes	
24	A	16 Sunday after Trinity.　*Sun apparently riſes and*	6 A 6	
25	M	*[ſets at 6, allowing for Refraction*	6	34
26	T	St. *Cyprian,* Abp. of *Carth.* and Mart.	7	9
27	W	Day decreaſed 4 hours 56 minutes	7	50
28	T	*Sheriffs* of *LONDON* Sworn. Day br. 4h. 11m.	8	39
29	F	S. *Michael* and all An. L. Mayor of *Lon.* Elected	9	38
30	S	S. *Jer.* Pr. Con. & D. *Sun* ri. 6h. 11m. ſets 5, 48	10	46

THE third of *September* the fourteenth is nam'd,
For which, *Britiſh* Annals will ever be fam'd ;
For by *Wiſdom* and *Art* to the Houſe made appear,
The *Sun* was reduc'd to attend on the *Year* ;
His *Julian* Vagaries long Time has he known ;
But has now got a new-bridal *Year* of his own.

in Oxfordshire, where the son of George Parker, the astronomer who made the speech that ignited Stanhope, was standing for Parliament. This election was depicted in a famous set of etchings by William Hogarth (1697–1764). In one of the etchings a banquet is being held by two Whig candidates—one of them is "Sir Commodity Taxem"— for their supporters. Everyone is reveling, with numerous small scenes showing people eating, a doctor tending to an injured man, musicians playing, and a man being struck in the head by a brick tossed by parading Tories. Lying on the floor at the feet of the wounded man is a poster: GIVE US BACK OUR 11 DAYS.

Other protesters shouted out a popular anti-reform ditty:

> *In seventeen hundred and fifty-three*
> *The style it was changed to Popery.*

In Bristol, riots over the reform apparently ended up with people killed. On January 6, 1753, which would have been the day after Christmas Day under the Old Style, a period journal reports:

> Yesterday being Old Christmas Day, the same was obstinately observed by our country people in general, so that (being market day according to the order of our magistrates) there were but a few at market, who embraced the opportunity of raising their butter to 9d. or 10d. per pound.

Also in Bristol a certain John Latimer reports that the Glastonbury thorn, which blossomed every year exactly on Christmas Day, "contemptuously ignored the new style" when it "burst into blossom on the 5th January, thus indicating that Old Christmas Day should alone be observed, in spite of an irreligious legislature."

In the City of London, bankers protested the reform and the confusion it caused for their industry by refusing to pay taxes on the usual date of March 25, 1753. They paid up 11 days later, on April 5, which remains tax day in Britain.

In a lighter vein, a correspondent wrote a letter to *The Inspector* that was published in the September 1752 issue of the popular *Gentleman's Magazine:*

> Mr. Inspector,
>
> I write to you in the greatest perplexity, I desire you'll find some way of setting my affair to rights; or I believe I shall run mad, and break my heart into the bargain. How is all this? I desire to know plainly and truly! I went to bed last night, it was Wednesday Sept. 2, and the first thing I cast my eye upon this morning at the top of your paper, was Thursday, Sept. 14. I did not go to bed till between one and two: Have I slept away 11 days in 7 hours, or how is it? For my part I don't find I'm any more refresh'd than after a common night's sleep.
>
> They tell me there's an act of parliament for this. With due reverence be it spoken, I have always thought there were very few things a British parliament could not do, but if I had been ask'd, I should have guess'd the annihilation of time was one of them!

Most people, however, did not seem too rattled by the change, with most diarists at the time simply mentioning the event with little comment. James Clegg, a 62-year-old minister and farmer living in Derbyshire, jotted down what he considered key events in his life for September 1752:

> 1. heavy rain all the forenoon, I was back home close at work writing my last will and was at home all day.
>
> 2. at home til afternoon then took a ride out into Chinley, visited at old William Bennets and at John Moults at Nase and returnd safe Blessed be God.
>
> 14. This day the use of the new Stile in numbring the days of the months commenceth and according to that computation, the last day of October will be my Birthday. I was at home til afternoon, we had an heavy shower of rain which raizd the water, after it was over I went up to Chappel on business and returned home in good time.

Newspapers also noted the change, but little more. None reported on the riots or other problems, since this was not yet part of what constituted a duty or practice of the general press. The *General Advertiser* of London printed excerpts from the act in its September 2, 1752 (Old Style) edition. The next day, September 14, was marked in the paper with a simple N.S. after the date. Otherwise the paper ran its usual mix of news from world capitals, shipping notices, stock quotes, and advertisements. The latter included on this first day of the new calendar an announcement that "The Evening Entertainments" at Spring Gardens, Vauxhall, "will end this evening, the 14th of September, N.S." There also was to be a violin concert that night at Islington, a sale of ten barges at Billingsgate at noon the following Tuesday, and a meeting of the governors of the Small-Pox Hospital on the 20th of September. High water at London Bridge was at 5:28 P.M.

Across the Atlantic in the British colonies, Benjamin Franklin's *Poor Richard's Almanac,* published in Quaker Philadelphia, noted:

> At the Yearly Meeting of the People called Quakers . . . since the Passing of this Act, it was agreed to recommend to the Friends a Conformity thereto, both in omitting the eleven Days of September . . . and beginning the Year hereafter on the first Day of the Month called January.

The author of this notice, R. Saunders, then extended a wish to his readers "that this New Year (which is indeed a New Year, such an one as we never saw before, and shall never see again) may be a happy Year."

In this same almanac Franklin himself, then 46, jauntily told his readers:

> Be not astonished, nor look with scorn, dear reader, at such a deduction of days, nor regret as for the loss of so much time, but take this for your consolation, that your expenses will appear lighter and your mind be more at ease. And what an indulgence is here, for those who love their pillow to lie down in peace on the second of this month and not perhaps awake till the morning of the fourteenth.

A number of colonial newspapers—including *The Boston Weekly News-Letter, The Carolina Gazette,* and *The New York Evening Post*—noted the arrival of the New Style but say little more.

Britain was not the last country to change in Europe. Sweden changed the next year, in 1753. Then there is a long gap, with the heavily Greek Orthodox countries in the Balkans waiting until the early twentieth century. Bulgaria made the switch in either 1912, 1915, or March of 1916, depending on which source one believes. Latvia, Lithuania, and Estonia converted around 1915, during the German occupation; Romania and Yugoslavia made the change in 1919. Russia waited until 1918, after the Bolshevik Revolution, but had to drop 13 days—February 1–13—to make up for the accumulation of days by which the Julian calendar was in error 336 years after the Gregorian reform. Greece did not reform its civil calendar until 1924.

Countries and people outside of Europe mostly had no reaction to the new calendar in the decades and centuries following 1582—the exception being in the Americas, where the reform was imposed by Spain and Portugal on those people they had conquered. These included the Aztecs, Incas, and Mayas, whose brilliant work in astronomy and calendars was already mostly forgotten and expunged by the Europeans, though to this day isolated groups of Mayans, for one, continue to use their ancient calendar. Later, Britain, France, the United States, and other colonial powers imposed their calendar on Indian tribes in the Western Hemisphere.

In Asia, the Japanese adopted the Gregorian calendar in 1873, during the Westernization period of the Meiji emperors. Most countries and peoples on this continent and in Africa preferred to keep their own calendars unless forced to change by European colonial powers. Many continue to use traditional calendars for religious and cultural events.

China resisted until 1912, though the Gregorian calendar did not take hold throughout that country until the victory of the Communists in 1949. On October 1 of that year a triumphant Mao Zedong climbed up onto a stand atop the Gate of Heavenly Peace, the main entrance to the imperial palace in Beijing. He then ordered that Beijing be henceforth the capital of China, that the red flag with a large gold star and four small stars be the official flag of China, and that the year be in accord with the Gregorian calendar.

But by then this calendar, launched 2,000 years earlier by Julius Caesar and modified 1,600 years later by a lackluster pope, had become the world's calendar: a code for measuring time that today all but the most isolated peoples use as the global standard for measuring time. This is despite its odd quirks and the twists of history that produced it, following an improbable timeline of its own from Sumer and Babylon to Rome, from Gupta India and the Islamic east to a gradually reawakening of Europe, the Renaissance, Lord Chesterfield's England, and beyond.

Today the quest continues in the age of atomic time—which takes us, at last, to Building 78 at the U.S. Naval Observatory in Washington, D.C., where time is now measured not by watching the moon and sun, or with a sundial, water clock, pendulum, wound-up spring, or quartz crystal, but with a tiny mass of a rare element called cesium.

15

Living on Atomic Time

But time is too large, it refuses to let itself be filled up.
—Jean-Paul Sartre

I am standing in front of the master clock.

It resides in a small bunker-like structure, on top of a grassy knoll. Here the output of some 50 individual atomic clocks converge into a bank of computers behind a thick pane of glass at the U.S. Naval Observatory. Smack in the middle of the panels and pulsating lights is a digital read out ticking past in bright red numbers: hours, minutes, and seconds. This is literally the pulse of North America in this age of atomic time. It also contributes to a larger system that keeps time for the entire world, accurate within a billionth of a second a year, or .0000000000114079 of a year.

Except that official time is no longer really measured this way, using antiquated seconds and years. Since 1972, when the atomic net went online, the Coordinated Universal Time—UTC—has been measured not by the motion of the earth in space but by the oscillations at the atomic level of a rare, soft, bluish-gray metal called *cesium*.

Apparently every atom oscillates—something I was unaware of before I visited the Naval Observatory. But before anyone gets alarmed, you should realize that all matter absorbs and emits a certain amount of energy, and that this happens in some elements with extraordinary regularity—absorb, emit, absorb, emit, absorb, emit—a process not unlike the steady swing of a pendulum, and which can be picked up by instruments as a steady frequency.

In 1967 the rate of cesium's pulse was calibrated to 9,192,631,770 oscillations per second. This is now the official measurement of world time, replacing the old standard based on the earth's rotation and orbit, which had used as its base number a second equal to 1/31556925.9747 of a year. This means that under this new regime of cesium, the year is no longer officially measured as 365.242199 days, but as 290,091,200,500,000,000 oscillations of Cs, give or take an oscillation or two.

What this means is that we humans have fulfilled the dream of Caesar, Aryabhata, al-Khwarizmi, Bacon, Clavius, and so many others: by creating a device at last that can measure a true and accurate year.

But alas, this is not the end of our story.

As we know, the earth wobbles and wiggles, causing random fluctuations in the earth's rotation. Which is why the master clock is *too* accurate, and must be periodically recalibrated. This is done by adding or subtracting leap seconds to compensate for the actual motion of the earth. Otherwise the master clock would gradually fall out of step with earth time, eventually rendering the atomic time-grid as erroneous for the nanosecond crowd as the Julian calendar became for those who could not tolerate drifts of minutes a year. Since 1972, leap seconds have been added almost every year. So far, none have been subtracted.

Think of the irony. After a millennia of struggle to come up with a true and accurate year, we have actually overshot the mark. For in the end the earth itself is not entirely accurate—a fact that would have astonished Roger Bacon and many an astronomer and time reckoner who fought to objectify time by using nature as their standard, as represented by the motions of the earth, planets, and stars. It seemed— well, more natural to them than a year devised by a church, emperor, parliament, or even a newfangled mechanical clock, which had to be rewound and reset and often ran fast or slow.

So we are left with yet another gap, one between atomic time and earth time, which fluctuates according to the whims of nature, if ever so slightly. Even with sophisticated modern instruments, time reckoners today can only watch and record the earth's bobs and dips as it drops a nanosecond here and gains two or three there. But they remain as helpless to predict the size of this gap at any moment as Bede was in the eighth century with *his* gap: between what he observed was the length of the year according to his sundial, and what the Church's calendar said the year should be. Indeed in his day the same problem existed with sacred time being too perfect compared to earth time, inasmuch as people believed it was perfect: though in this case the perfection came from God, not of cesium.

Which leaves anyone living by the clock or the calendar trapped in a conundrum of our own making, between our seemingly genetic compulsion for order and perfection, and the plain reality that nothing is perfect, particularly nature—something we relearn every hurricane season, and whenever the latest Theory of Everything falls short.

Further complicating matters is the fact that our little planet offers not one or two but several "years" to measure, each slightly different. I have several times mentioned the Sidereal Year: the year as measured by the time it takes for the earth to orbit the sun. And of course the Tropical Year, defined as a year measured from one March equinox to the next, though this is not entirely accurate in modern astronomy, if one gets picky. Officially, a Tropical Year is the time interval it takes for the earth to make a full orbit of the sun, using as the starting and stopping point the vernal equinox. This is slightly different than the value for the equinoctial year, since the earth's rotation is slowing ever so slightly over time. This means that the point where the equinox started in a given year in relation to the sun is not going to be exactly the same point a year later due to the slowing, and to other planetary fluctuations.

If this is not numbing enough, we also have the year as measured from one June solstice to the next, from one September equinox to the next, and so forth—all of which offer up minutely varying values for the length of the year that would have left the heads spinning of Sosigenes, Bede, Roger Bacon, and all the rest.

In the spirit of full disclosure, for those who are interested, I list below the various "years" and their values for the year 2000.*

Year	Mean Time Interval, Year 2000 (in days)
Sidereal	365.2564 days
Tropical	365.24219 days
Between two March equinoxes	365.24237 days
Between two June solstices	365.24162 days
Between two September equinoxes	365.24201 days
Between two December solstices	365.24274 days

Our calendar year is linked to the year as measured between two March equinoxes, as originally established by Caesar and Sosigenes. Pope Gregory's correction in 1582 brought our calendar year within 26 seconds of the equinoctial year, where it remains today.

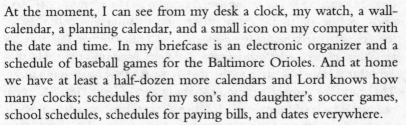

At the moment, I can see from my desk a clock, my watch, a wall-calendar, a planning calendar, and a small icon on my computer with the date and time. In my briefcase is an electronic organizer and a schedule of baseball games for the Baltimore Orioles. And at home we have at least a half-dozen more calendars and Lord knows how many clocks; schedules for my son's and daughter's soccer games, school schedules, schedules for paying bills, and dates everywhere.

This begs the question: Why do we need to measure a picosecond when I cannot even keep track of what I am doing day to day?

---

*These values were determined by astronomer Jean Meeus in 1992, except for the sidereal year.

I posed this to historian Steve Dick at the U.S. Naval Observatory. An affable, quiet man with short brown hair and a well-kept mustache, he laughed and said that *everyone* asks this. "You would be surprised at how many uses there are," he said, starting with navigation—which was the original impetus to start a national time synchronization system here at the Naval Observatory in the nineteenth century.

According to him a billionth of a second translates into the space of about one foot for navigation, which can be significant if you are a pilot at night in the fog trying to land a Boeing 747 on a runway or an F-14 Tomcat onto an aircraft carrier. These sorts of minute measurements are critical for synchronizing television feeds, bouncing signals off satellites, calculating bank transfers, transmitting everything from e-mail to sonar signals in a submarine, and keeping "smart" missiles on course so they slam into an enemy's chemical weapons complex instead of the middle of a populated neighborhood. Hikers in the wilderness are using the master clock to find trails down to a foot or two in accuracy by using handheld GPS (Global Positioning System) locators. These locators cost as little as $250, and work by simply holding the device up to the sky and waiting for it to link up with three or more satellites. Once contact is made, the locator flashes an exact location in degrees, minutes, and seconds.

But wait—trying to determine a true year gets much more mind-boggling than this. For when we get down to the world of nanoseconds time begins to change in other ways that must be compensated for. Time in fact begins to *warp* or *bend* noticeably at this level of precision under certain situations, as Albert Einstein noted. He theorized that time is relative to the speed one is traveling through space. That time for someone moving at the speed of light (186,282 miles per second) would move much more slowly than someone moving, say, on the earth as it hurtles through space around the sun at about 20 miles a second.

This was proved in 1971 when two scientists borrowed four atomic clocks from the Naval Observatory and flew them east and west around the globe in jet airliners. Comparing the nanoseconds of these voyages off the earth to atomic clocks on the ground showed that people flying in the jets at less than one-millionth the speed of light experienced a slowing of time equal to 59 nanoseconds going east and 273 nanoseconds going west—the difference caused by the fact that the earth is rotating to the east.

What does this mean for measuring the year? For one thing, it means that every time someone flies their "year" grows by a few billionths of a second: which is entirely meaningless, since the earth's fluctuations affect the length of the year in the range of a thousandth of a second. But who knows. It may matter a great deal if humans learn to travel at great speeds through space, where a "year" in a spaceship moving at 186,000 miles per second would last far longer than 365.242199 earth days.

Lost in this expanding universe of cesium, nanoseconds, warps, and recalibrations is the lowly calendar, with its twelve months and 365 little boxes (366 in a leap year): a device for measuring time that does not oscillate, bend time, or have anything to do with the electromagnetic spectrum. Invented in its present form over two thousand years ago and corrected only once four centuries ago, it is old enough to be an artifact in a museum.

Except that it remains vital to everyone.

Not that it is even close to perfect. There is a host of minor flaws that annoy people, and are forever keeping a small, but vibrant group of would-be reformers hoping to get a new and improved calendar named after them. These flaws include:

- The divisions of the year—the month, the quarter, the half year— are of unequal length. This is most unpleasant for anyone trying to run a business, pay taxes, or gather statistics.
- The days of the week drift each year, with each new year starting

the day after the previous year, or two days after when following a leap year. Because of the leap year, this drift runs a cycle that repeats itself only after 28 years. This makes it difficult to fix annual dates, since they keep moving in the week. The position of the weeks also moves each year within months and quarters.

- The Gregorian calendar remains in error, running fast against the true year by about 25.96 seconds a year. Since 1582, this has accumulated to about 2 hours, 59 minutes, and 12 seconds and will equal an entire day about 72 generations from now—in 4909—assuming humans are still here, and are still using the calendar named for a pope who died 3,330 years earlier.

- The "era" we use to number our years—initially called the "Christian Era" and now the "Common Era"—remains confusing because there is no year zero. This means that technically century-years come in the −01 slot, not −00, and millennium years happen in −001, not −000. But people prefer to celebrate the beginning of, say, the twentieth century as 1900, and the coming millennium in the year 2000, not 2001. Others complain about the awkwardness of an A.D. and B.C. timeline with "positive" and "negative" dates.

Over the years, attempts have been made to fix these pesky little problems. One of the most intriguing of these was the French Revolutionary Calendar—the "Calendar of Reason." It did nothing to correct the 25.96 seconds error, which the revolutionaries were probably unaware of. But they did fix other calendric conundrums in their zeal to expunge the old order in the same way Caesar, Constantine, and so many others did.

In this case the French Jacobins simply threw out the Gregorian calendar and replaced it with their own—which happened to be far more uniform and convenient. Launched in 1792—the revolutionary Year One—this new calendar had uniform months of 30 days each, tacking on the extra 5 (or 6) days at the end. These were reserved for holidays called *Virtue, Genius, Labor, Opinion,* and *Recompense.*★ In-

---

★The Maya and Aztecs used a similar arrangement; so did the Egyptians.

stead of gods and emperors it used names for the months: *Nivose* for snowing months, *Pluviose* for Rainy Month, *Thermidor* for Heat Month, and *Brumaire* for Foggy Month.* Weeks were 10 days long, with three weeks per month. Days were likewise divided in a decimal arrangement into 10 hours each of 100 minutes, with every minute containing 100 seconds.

The Calendar of Reason was a great improvement, but it lasted only until 1806, when Napoleon quietly reinstated the Gregorian system. The experiment did produce a number of curious watches and clocks with ten hours, and minutes divided up into decimals; and numerous books published with single-digit years.

More recently, reform efforts have centered on trying to tinker with the Gregorian calendar, the most popular being the proposed World Calendar, sometimes known as the Universal Calendar. This would restore Caesar's original distribution of the 12 months as alternating between 30 and 31 days: which Augustus and the Roman Senate altered in A.D. 8 to give Augustus the same number of days in August as Caesar had in July. The World Calendar would start each year and each quarter on a Sunday. And each month would always start on the same day. Leap days would simply be an extra day, not attached to a month. One plan was to declare this special day "World's Day."

Supporters of the World Calendar have several times tried to get the United Nations to endorse this reform—coming close in 1961, which started on a Sunday. In 1954 the Vatican endorsed the World Calendar; it was even introduced in the U.S. Congress. But it never caught on.

Other proposals continue to come and go, including a calendar of 13 months with 28 days each, with an extra day (or two) tacked on as special days. This was the favorite choice of the 1929 National Committee on Calendar Simplification for the United States, chaired by George Eastman of Eastman Kodak.† One of the many more re-

---

*Wags in Britain made fun of these French months, calling them: *wheezy, sneezy, freezy, slippy, drippy, nippy,* and so forth.

†This committee was convened at the request of the League of Nations, which or-

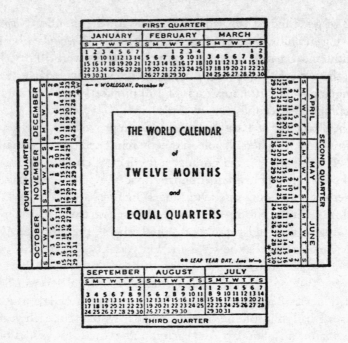

cent ideas is something I saw on the Internet. Called "The Goddess Lunar Calendar," its proponents advocate a 25-month calendar of alternating 29 and 30 days; with each of the months named after a female deity: Artemis, Bast, Cybele, and Gaia, to name a few.

Fixing the 25.96 seconds error is much simpler. Indeed, proposals have been made to slip in a leap-*millennium* rule, which would cancel out the Gregorian leap-century rule by eliminating a leap day on millennial years such as the year 2000. This would make the "modified" Gregorian calendar accurate to within a day every 3,323 years. Undoubtedly, this fix will become official sometime in the next millenium, or the one after, if in fact the world is still using Gregory's calendar.

---

ganized a worldwide effort in 1928–29 to simplify the calendar, without success. The U.S. committee was composed of dozens of prominent Americans, including Eastman, Henry Ford, Adolph Ochs, Gilbert Grosvenor, and George P. Putnam.

As for the problem with no year zero, I know of no plans to make a correction: which at the very least would involve changing every history book dealing with dates before the year one A.D. In calendar circles new ideas come and go—with proposals floating around suggesting a new chronological system that would start with a year one according to formulas and at various moments in history.

Just the other day a calendar group on the Internet had a brief discussion that began by someone noting that the September equinox in 1997 would be the Year 6000 in the time line established by the Irish prelate and scholar James Ussher (1581–1656). He proposed that God created the world on the 23rd of October, 4004 B.C. Another participant fired back that under the Byzantine calendar—whatever that is—the year 7506 had just begun. "The reason it holds special interest for me is that it starts earlier than any other calendar I have seen," writes this calendar-aficionado. "If we used that date, most of recorded history would have a positive date, and it would eliminate the need for B.C."

Another calendar listserver member replied:

"A much simpler solution would be to just add 10,000 to the current year number. [Then] it would be very easy to observe that, e.g., in 2011 we will commemorate the 2500th anniversary of the run at Marathon." He also points out the ludicrous practice of a B.C. calendar that counts years backward, but starts each of these negative years on January 1, after which they run forward through the days and months as if they were on the "positive" side of the B.C./A.D. split.

This observation was followed by someone mentioning a calendar proposed several years ago called the Holocene Calendar that would use the end of the last ice age as its starting point, some 12,000 years ago. Which prompted a flurry of other responses and ideas in a debate that at least in this small corner of cyberspace is not going away.

Meanwhile, as the cesium atoms in the master clock continue to oscillate, and the earth wobbles and slows ever so slightly, most of us carry on as people have since we first became aware of time—whether we live by the Gregorian, Holocene, Zoroastrian, Hebrew, Babylonian, Nuer, Moslem, or Goddess Lunar calendar. We take in stride a calendar used by most of the world that is flawed, but endures, largely because it works just fine for most of us, and it is what we are used to.

As I watched the red numbers flash past on the master clock in Building 78, I was out of time myself. My calendar for that date, September 18, said I needed to be uptown at an appointment at 11:30 A.M., in a mere 8,273,368,593,000,000,000,000 oscillations (or so) of cesium. That's roughly 15 minutes in old-fashioned earth time. Though in any time, except perhaps Einstein's warped time, I was going to be late, which made me want to swear at the little square box in my date book so crammed with things to do that I was going to spend the whole day being late.

Which brought new meaning to the words of Sartre, who I think got it backward when he said "But time is too large, it refuses to let itself be filled up."

He was talking about clock time: the endless cycles of seconds, minutes, and hours that go on and on. By contrast calendar time is all about those little boxes of days strung out one after another, all squeezed into a finite and artificial time span of our own making. After all it was we humans who invented this thing that is both a miraculous tool and a cage of finite moments that keep us forever running about, trying to make the best of the short time we have been alotted. At least those of us in the West who are more obsessed than anyone else with counting oscillations and cramming little boxes with things to do.

There are moments when I am hopelessly late, or cannot possibly fit anything else into my schedule, when I sigh and wish that Cro-Magnon man 13,000 years ago in the Dordogne Valley had set aside his eagle bone unfinished and had gone to bed. Or gone for a long walk under the Paleolithic sky. Or gone to play with his little Cro-Magnon children. Of course, this would have only delayed the inevitable as some other fur-clad hominid took up the task of carving notches and counting phases of the moon, launching humanity on its strange, epic quest.

And now I have to go, because I am out of time.

TIME LINE
A Chronology of Events

The Calendar

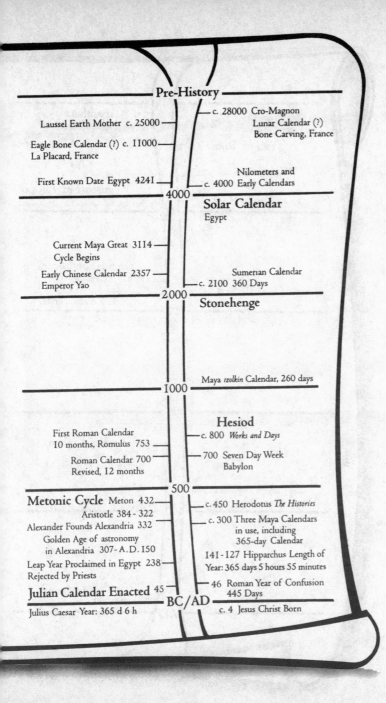

Pre-History

c. 28000 Cro-Magnon
Lunar Calendar (?)
Bone Carving, France

Laussel Earth Mother c. 25000

Eagle Bone Calendar (?) c. 11000
La Placard, France

Nilometers and
First Known Date Egypt 4241   c. 4000 Early Calendars

4000

Solar Calendar
Egypt

Current Maya Great 3114
Cycle Begins

Early Chinese Calendar 2357   Sumerian Calendar
Emperor Yao   c. 2100 360 Days

2000

Stonehenge

Maya *tzolkin* Calendar, 260 days

1000

Hesiod
c. 800 *Works and Days*

First Roman Calendar
10 months, Romulus 753

700 Seven Day Week
Roman Calendar 700   Babylon
Revised, 12 months

500

Metonic Cycle Meton 432
Aristotle 384 - 322   c. 450 Herodotus *The Histories*
Alexander Founds Alexandria 332   c. 300 Three Maya Calendars
Golden Age of astronomy   in use, including
in Alexandria 307 - A.D. 150   365-day Calendar
                                       141 - 127 Hipparchus Length of
Leap Year Proclaimed in Egypt 238   Year: 365 days 5 hours 55 minutes
Rejected by Priests
                                       46 Roman Year of Confusion
Julian Calendar Enacted 45   445 Days

BC/AD

Julius Caesar Year: 365 d 6 h   c. 4 Jesus Christ Born

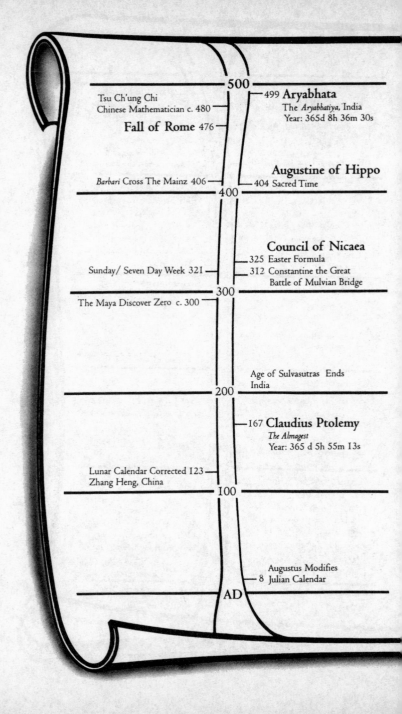

500

Tsu Ch'ung Chi
Chinese Mathematician c. 480

Fall of Rome 476

— 499 **Aryabhata**
The *Aryabhatiya*, India
Year: 365d 8h 36m 30s

Augustine of Hippo

Barbari Cross The Mainz 406
— 404 Sacred Time

400

Council of Nicaea

— 325 Easter Formula
— 312 Constantine the Great
Battle of Mulvian Bridge

Sunday/ Seven Day Week 321

300

The Maya Discover Zero c. 300

Age of Sulvasutras Ends
India

200

— 167 **Claudius Ptolemy**
The Almagest
Year: 365 d 5h 55m 13s

Lunar Calendar Corrected 123
Zhang Heng, China

100

Augustus Modifies
— 8 Julian Calendar

AD

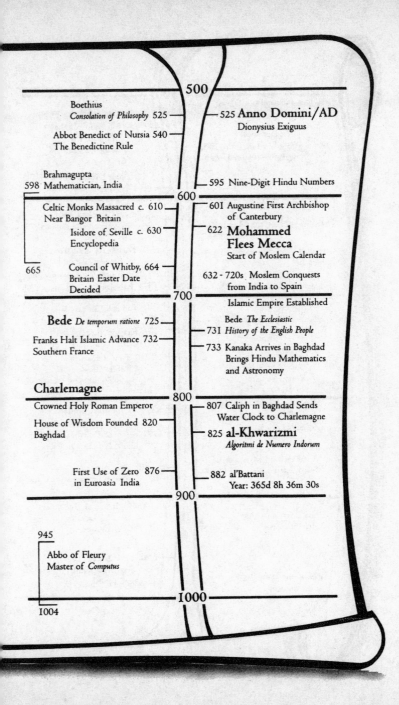

500

Boethius
Consolation of Philosophy 525

— 525 **Anno Domini/AD**
Dionysius Exiguus

Abbot Benedict of Nursia 540 —
The Benedictine Rule

Brahmagupta
598 Mathematician, India

— 595 Nine-Digit Hindu Numbers

600

Celtic Monks Massacred c. 610 —
Near Bangor Britain

— 601 Augustine First Archbishop
of Canterbury

Isidore of Seville c. 630 —
Encyclopedia

622 **Mohammed
Flees Mecca**
Start of Moslem Calendar

665

Council of Whitby, 664 —
Britain Easter Date
Decided

632 - 720s Moslem Conquests
from India to Spain

700

Islamic Empire Established

Bede *De temporum ratione* 725 —

Bede *The Ecclesiastic*
— 731 *History of the English People*

Franks Halt Islamic Advance 732 —
Southern France

— 733 Kanaka Arrives in Baghdad
Brings Hindu Mathematics
and Astronomy

Charlemagne

800

Crowned Holy Roman Emperor

— 807 Caliph in Baghdad Sends
Water Clock to Charlemagne

House of Wisdom Founded 820 —
Baghdad

— 825 **al-Khwarizmi**
Algoritmi de Numero Indorum

First Use of Zero 876 —
in Euroasia India

— 882 al'Battani
Year: 365d 8h 36m 30s

900

945

Abbo of Fleury
Master of *Computus*

1000

1004

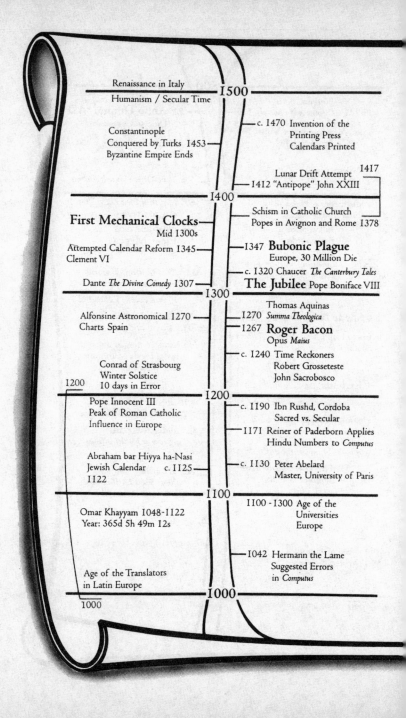

Renaissance in Italy
Humanism / Secular Time

1500

c. 1470 Invention of the
Printing Press
Calendars Printed

Constantinople
Conquered by Turks 1453
Byzantine Empire Ends

Lunar Drift Attempt   1417
1412 "Antipope" John XXIII

1400

Schism in Catholic Church
Popes in Avignon and Rome 1378

First Mechanical Clocks
Mid 1300s

1347 **Bubonic Plague**
Europe, 30 Million Die

Attempted Calendar Reform 1345
Clement VI

c. 1320 Chaucer *The Canterbury Tales*

Dante *The Divine Comedy* 1307

The Jubilee Pope Boniface VIII

1300

Thomas Aquinas

Alfonsine Astronomical 1270
Charts Spain

1270 *Summa Theologica*

1267 **Roger Bacon**
Opus *Maius*

c. 1240 Time Reckoners
Robert Grosseteste
John Sacrobosco

Conrad of Strasbourg
Winter Solstice
1200   10 days in Error

1200

Pope Innocent III
Peak of Roman Catholic
Influence in Europe

c. 1190 Ibn Rushd, Cordoba
Sacred vs. Secular

1171 Reiner of Paderborn Applies
Hindu Numbers to *Computus*

Abraham bar Hiyya ha-Nasi
Jewish Calendar   c. 1125
1122

c. 1130 Peter Abelard
Master, University of Paris

1100

Omar Khayyam 1048-1122
Year: 365d 5h 49m 12s

1100 - 1300 Age of the
Universities
Europe

1042 Hermann the Lame
Suggested Errors
in *Computus*

Age of the Translators
in Latin Europe

1000

1000

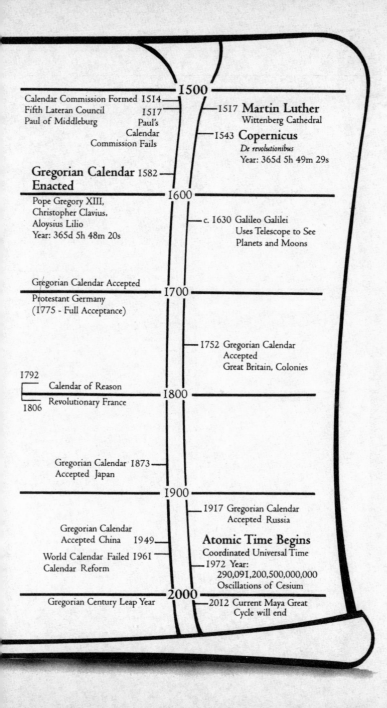

— **1500** —

Calendar Commission Formed 1514 —
Fifth Lateran Council   1517 —
Paul of Middleburg   Paul's
                          Calendar
                          Commission Fails

— 1517 **Martin Luther**
        Wittenberg Cathedral

— 1543 **Copernicus**
        De revolutionibus
        Year: 365d 5h 49m 29s

Gregorian Calendar 1582 —
Enacted

— **1600** —

Pope Gregory XIII,
Christopher Clavius,
Aloysius Lilio
Year: 365d 5h 48m 20s

— c. 1630 Galileo Galilei
          Uses Telescope to See
          Planets and Moons

Gregorian Calendar Accepted

— **1700** —

Protestant Germany
(1775 - Full Acceptance)

— 1752 Gregorian Calendar
        Accepted
        Great Britain, Colonies

1792
⌐ Calendar of Reason
└─ Revolutionary France
1806

— **1800** —

Gregorian Calendar 1873 —
Accepted Japan

— **1900** —

— 1917 Gregorian Calendar
        Accepted Russia

Gregorian Calendar
Accepted China   1949 —

Atomic Time Begins
Coordinated Universal Time
— 1972 Year:
        290,091,200,500,000,000
        Oscillations of Cesium

World Calendar Failed 1961 —
Calendar Reform

— **2000** —

Gregorian Century Leap Year

— 2012 Current Maya Great
        Cycle will end

Illustrations

Bibliographic Notes

I wrote this book as a storyteller fascinated by the unusual and unexpected tale of how the calendar used by most of the world came to be. I make no claims to scholarly expertise in the far-ranging fields of time reckoning, astronomy, mathematics, the philosophy of time, theology, or history. I have done my best to meticulously and accurately research what was necessary in each of these fields to write this story. I have consulted with experts who generously gave their time and generally agreed with my interpretations, or helped me to correct them. Obviously any mistakes or misinterpretations are my own.

The following are highlights of sources used in writing this book. I conservatively consulted many hundreds of sources, both primary and secondary. On general historic topics I tended to consult a number of secondary sources, checking them against one another, and asking my expert readers to review the material. On calendar issues I used primary sources where possible. I worked extensively in the Library of Congress and visited and worked in the British Library in London and Vatican Library and archives in Rome.

I found surprisingly few recent books written on the calendar, though an excellent and lively work on the history and meaning of time was recently published by the astroarchaeologist Anthony F. Aveni, called *Empires of Time: Calendars, Clocks and Cultures* (Kodansha International, 1995). I also found particularly helpful J.T. Fraser's *Time: The Familiar Stranger* (The University of Massachusetts Press,

1987); Margo Westrheim's thin but highly informative volume *Calendars of the World* (Oneworld, 1993); and P.W. Wilson's classic *The Romance of the Calendar* (Norton, 1937). Also *The Book of Calendars*, ed. by Frank Parise (Facts on File, 1982).

Most indispensable of all the general works was a paperbound collection of essays I happened to find at the Vatican bookstore in Rome: *Gregorian Reform of the Calendar: Proceedings of the Vatican Conference to Commemorate Its 400th Anniversary 1582–1982*, edited by G.V. Coyne, M.A. Hoskin and O. Pedersen (Specola Vaticana, 1983). This collection includes offerings from calendar experts from around the world who detail all aspects of the Gregorian reform, the history of the Catholic ecclesiastic calendar, the reaction to the 1582 reform, and the currect status of the calendar.

For ideas and general information on the history of time and science I consulted Daniel J. Boorstin's *The Discoverers: A History of Man's Search to Know His World and Himself* (Vintage, 1983) and an assortment of encyclopedias: *The World Book Encyclopedia* (1995); *The New Catholic Encyclopedia* (1967); *A History of Technology*, ed. by C. Singer, et al. (Clarendon Press, 1954); and the *Dictionary of Scientific Biography*, ed. by C.C. Gillispie (1970–1980). Atlases and general historic works include Norman Davies' *Europe: A History* (Oxford University Press, 1996); Colin McEvedy's *The Penguin Atlas of Medieval History* (1967) and *The Penguin Atlas of Medieval History* (1969); and *The Times Concise Atlas of World History*, ed. by Geoffrey Barraclough (Hammond, 1982). And the indispensable *Webster's New Biographical Dictionary* (1983); *Webster's New Geographical Dictionary* (1984); and *Webster's New Universal Unabridged Dictionary* (1972).

Internet sites included the New Advent Catholic Supersite, *http://www.knight.org/advent/cathen/*; CalendarLand, *http://website.juneau.com/home/janice/calendarland/*, a general information site on calendars past and present from around the world; and Britannica Online, *http://www.eb.com/*. I also used numerous sites on topics ranging from mathematics to descriptions of cities and countries; and the philosophy of time to the Black Plague.

On early calendars and societies I used Aveni's *Empires of Time*; Alexander Marshack's classic: *The Roots of Civilization* (McGraw-Hill, 1972); Dr. Marshack also was kind enough to send me a number of articles updating his work. I consulted Michael Coe's *The Maya* (Thames and Hudson, 1993); John Phelps, *The Prehistoric Solor Calendar* (Johns Hopkins Press, 1955); *Archaeoastronomy in the New World*, ed. by Anthony Aveni (Cambridge University Press, 1982); G.S. Hawkins, *Stonehenge Decoded* (Delta Dell, 1965), and C. Chippindale, *Stonehenge Complete* (Cornell University Press, 1983).

On the general history of science, time and the calendar: Arno Borst, *The Ordering of Time: From the Ancient Computus to the Modern Computer* (Polity Press, 1993) and *Ancient Inventions*, ed. by Peter James and Nick Thorpe (Ballantine Books, 1994). Also Gerhard Hohrn-van Rossum, *History of the Hour: Clocks and Modern Temporal Orders*, trans. by Thomas Dunlap (University of Chicago Press, 1996). On philosophy, *A History of Philosophy*, by Frederick Copleston (Doubleday, New York, 1985).

On the history of astronomy, I drew on A. Pannekoek, *A History of Astronomy* (Dover, New York, 1961); Hugh Thurston, *Early Astronomy* (Springer-Verlag, New York, 1994); and *The Cambridge Illustrated History of Astronomy*, ed. by Michael Hoskin (1997). For ancient Alexandria I consulted Kenneth Heuer, *City of Stargazers* (Scribner's, 1972). On general astronomy, Fred L. Whipple, *Orbiting the Sun, Planets and Satellites of the Solar System* (Harvard University Press, 1981) and Jean Meeus, *Astronomical Tables of the Sun, Moon and Planets* (Willmann-Bell, Richmond, Virginia, 1983). Also by Jean Meeus and Denis Savoie, "The History of the Tropical Year," *Journal British Astronomical Association*, 102, 1, 1992: 40–2. On the history of mathematics, Carl B. Boyer, *A History of Mathematics* (John Wiley & Sons, 1991) and G.G. Joseph, *The Crest of the Peacock* (Penguin, 1992). On the science of time, Paul Davies, *About Time, Einstein's Unfinished Revolution* (Touchstone, 1995) and Stephen Hawking, *A Brief History of Time* (Bantam, 1988).

For the Roman calendar I used Agnes Kirsopp Michels' *The Cal-*

endar of the Roman Republic (Princeton University Press, 1967); Van Johnson, *The Roman Origins of Our Calendar* (American Classical League, 1974); and Carole E. Newlands, *Playing with Time, Ovid and the Fasti* (Cornell University Press, 1995). On general Roman history, Edward Gibbon, *The Decline and Fall of the Roman Empire*; J.B. Bury, *History of the Later Roman Empire* (Dover, 1958); and *The Cambridge Ancient History*, vol. IX, ed. by J.A. Crook, et al. (Cambridge University Press, 1994).

On general medieval and Renaissance history I used Norman F. Cantor, *The Civilization of the Middle Ages* (HarperPerennial, 1993); Maurice Keen, *The Pelican History of Medieval Europe* (Penguin, 1988); and Eugene F. Rice, Jr., *The Foundations of Early Modern Europe, 1460–1559* (Norton, 1970). On science in the Middle Ages and the Renaissance: *Science in the Middle Ages*, ed. by David C. Lindberg (University of Chicago Press, 1978); Edward Grant, *The Foundations of Modern Science in the Middle Ages* (Cambridge University Press, 1996); and Alfred W. Crosby, *The Measure of Reality: Quantification and Western Society, 1250–1600* (Cambridge University Press, 1997). On the Church in the Middle Ages I used Margaret Deanesly, *A History of the Medieval Church, 590–1500* (Methuen & Co, 1969). I also used Jacques Le Goff, *Intellectuals in the Middle Ages*, trans. by Teresa Lavender Fagan (Blackwell, 1994) and Le Goff's *Medieval Civilization, 400–1500*, trans. by Julia Barrow (Blackwell, 1995). For primary sources I used the truly phenomenal "Internet Medieval Sourcebook," out of Fordham University, at *http://www.fordham.edu/halsall/sbook2.html*, which includes extensive offerings of complete and often hard to find original texts.

For India I consulted Romila Thapar, *A History of India* (Penguin Books, 1977); for the history of Islam and the Arab empire, Philip K. Hitti, *The Arabs, A Short History* (Regency Publishing, 1996). On the Moslem calendar I read G.S.P. Freeman-Grenville, *The Muslim and Christian Calendars* (Oxford University Press, 1963).

The Gregorian reform itself is cited in exhaustive detail in a number of sources already mentioned. I also used a number of primary sources, including the original bull issued by Gregory XIII, the *Compendium Novae Rationis Restituendi Kalendarium* issued by the pope's calendar commission, and other documents housed in the Vatican archives and in other libraries. I drew heavily from the *Gregorian Reform of the Calendar*, cited above.

To recount Britain's reform of the calendar I used a number of primary sources, including British state papers from the reign of Queen Elizabeth I, and for the 1750s; and several newspapers and pamphlets from England and the American colonies from 1751–1753. *The Gentleman's Magazine* also provides detailed and entertaining accounts of the reform effort in the 1580s and in 1752. See the issues from March 1751; April 1751; and September 1752. Also see an informative little booklet by H. Dagnall, "Give Us Back Our Eleven Days: An Account of the Change from the Old Style to the New Style Calendar in Great Britain in 1752," (published by the author, Queensbury, U.K., 1991).

For profiles of major characters I read each subject's original works, plus biographies, articles, and biographic citations from encyclopedias and biographical dictionaries. For Roger Bacon the secondary sources included Stewart C. Easton, *Roger Bacon and His Search for Universal Science* (Russel and Russel, 1971); Winthrop F. Woodruff, *Roger Bacon, A Biography* (James Clark & Co., 1938); and Lynn Thorndike, "The True Roger Bacon," *The American Historical Review*, vol. XXI no. 3, January and February, 1916. On Copernicus I read Angus Armitage's *The World of Copernicus* (E.P. Publishing, Ltd., 1972). On Lord Chesterfield: Colin Franklin, *Lord Chesterfield: His Character and Characters* (Scholar Press, 1993). On Christopher Clavius and sixteenth century Roman intellectual life: James M. Lattis, *Between Copernicus and Galileo: Christoph Clavius and the Collapse of Ptolemaic Cosmology* (University of Chicago Press, 1994). On John Dee: Richard Deacon, *John Dee: Scientist, Astrologer, and Secret Agent to Elizabeth I* (Frederick Muller, 1968) and William H. Sherman, *John Dee, The Politics of Reading and Writing in the English Renaissance* (University of Massachusetts Press, 1995). On Julius Caesar: J.F.C. Fuller, *Julius Caesar* (Da Capo

Press, 1965) and Christian Meier, *Caesar* (Basic Books, 1982). On Constantine: Michael Grant, *Constantine the Great* (Scribner's, 1993). On Bede: Charles Jones, *Bede, the Schools and the Computus* (Variorium, 1994).

Acknowledgments

Words cannot express my thanks to my family for putting up with me for months on end working late at night and on weekends to write this book: my beautiful and understanding wife Laura; my children Sander, Danielle, and Alex. My dad, who read the manuscript and was a great help and inspiration. And my mother, who has always been my most enthusiastic supporter.

A warm thanks to Stephen Power, an extraordinary editor who asked me to write this book, somehow knowing it would be a delight for me and a wondrous learning experience. To Mel Berger, who has always believed in me and encouraged me: he is the greatest agent I know of. Thanks to Marcie Posner, globetrotting agent at William Morris, and Claudia Cross. And also to Sue Warga, copy editor with no peer, and master of a thousand details.

Thanks to Polly Bart, an extraordinary researcher and friend; and my assistant Tanya Vlach, who leapt into the fray at the end to help get the book out the door.

A number of scholars and advisors helped me attempt to understand and get right the history and facts represented in this text. Thanks to my old friend Steve Vicchio and other expert readers: Anthony Aveni, Richard Landes, Tom Settle, Rick McCarty, and Steve Dick. I also appreciate the help of the librarians and researchers at the Library of Congress, the British Library in London, the Vatican Library in Rome, and the Milton S. Eisenhower Library at Johns Hopkins University in Baltimore. Also thanks to Richard Hansen, David Joyce, Clive Priddle, Brett Robertson, the U.S. Naval Observatory, and the Royal Observatory in Greenwich, England. And to Richard Harris, Tom Bettag, and the staff at ABC Nightline, and the staff in the Washington and Rome bureaus of ABC News.

Index

Page numbers in italics indicate illustrations